A Basic Concept of Civil Engineering

Sunder Narayan

ATLANTIC

PUBLISHERS & DISTRIBUTORS (P) LTD

Published by

ATLANTIC

PUBLISHERS & DISTRIBUTORS (P) LTD

7/22, Ansari Road, Darya Ganj, New Delhi-110002
Phones : +91-11-40775252, 40775214, 23273880, 23275880
Fax : +91-11-23285873
Web : www.atlanticbooks.com
E-mail : orders@atlanticbooks.com

Branch Office: Chennai
Phones : +91-44-48531784, 28291383
E-mail : chennai@atlanticbooks.com

Printed & bound in India by Atlantic Print Services

Preface

In my experience as a civil engineer under various capacities has made me realise how this field has become a secondary field. The importance of this field has been diminishing over the years. Like the story of blind men and the elephant, each person has understood and defined this field in their own way. Predominantly the understanding is that a civil engineer role is limited to that of a contractor who builds. The vastness of this field has not been propagated. The concept of who, what and why of being a civil engineer is rarely explored and explained to aspiring civil engineers. Engineer being an applied sciences, it is important that people know the applications of their learning.

My effort in this book has been to explain the concept rather than the details of the topics. I have tried my best to explore and explain what it is, the history of where it started, who laid the foundation, why it is being done and some basic principles being applied. My aim is to create an interest in this field who think this field is dull.

In my first chapter I have tried to educate the people on what is civil engineering and scope of a civil engineer. It will particularly help the students in understanding where they and where they want to go. In the subsequent chapters I have briefly explained the concepts and the process. These concepts will not only help a civil engineer but even architects and non-engineers who have a keen interest for a better understanding of this field. I have tried to connect past and present of various topics. In my final chapter I have talked about Temples. My effort is to educate the readers that the basic processes haven't changed and also create awareness that the people of ancient times were as modern as we are now.

To conclude, I would like to narrate a small story of Sage Bharadwaja. It is said that he wanted enough lifetime to master the whole of Vedas. The Gods appeared before him and created 3 mountains. They took 3 handfuls, one from each mountain and said, "Bharadwaja, the learning you have is like the handful of mud and the Vedas are like the three mountains. This doesn't mean you know very little. As much as it is important to learn, it is more important to spread what one has learnt."

Sunder Narayan

Acknowledgements

My belief is that any achievement is never an individual success and the success of an individual or an event is a combined effort. Though writing has been my passion, this would never have been possible without the people involved in this process.

I would like to thank my parents who were also my first guru and teacher. My father being a civil engineer himself, was my inspiration for my choice of civil engineering as a profession.

I would also like to thank the following people in giving their valuable inputs and making this a reality:

Dr. K.S. Satyanarayanan who was my project guide during my masters; my friends and colleagues, C.R.Sriraman, Ar. S.B. Raju, P. Magesh, and K.K. Revathi, for their invaluable inputs to the content of the book. I am particularly grateful to Raju for his help with the drawings.

And last but not least, I would like to deliver special thanks to my dear friend, Sundari, who has been inspiring me to become a life coach and then continuously prodding me to complete this venture.

Sunder Narayan

Contents

Abbreviations

IS Code	Indian Standard Code
PCC	Plain Cement Concrete
RCC	Reinforced Cement Concrete
Rpm	Revolutions per minute
ASTM	American Society for Testing and Materials
BSI	British Standard Institute
ISI	Indian Standard Institution
BIS	Bureau of Indian Standards
SBC	Safe Bearing Capacity
OPC	Ordinary Portland Cement
PPC	Portland Pozzolonic Cement
GIS	Geographical Information System
QS	Quantity Surveyor
QC	Quality Control
QA	Quality Assurance
CAD	Computer Aided Design
AAC	Autoclaved Aerated Concrete
pH	Potential of Hydrogen
CaCl	Calcium Chloride
HYSD	High Yield Strength Deformed
PVC	Poly Vinyl Chloride
UPVC	Unplasticised Poly Vinyl Chloride
NGL	Natural Ground Level
GFC	Good For Construction
BOQ	Bill Of Quantities

FIDIC	Fédération Internationale Des Ingénieurs-Conseils
ISO	International Standard Organization
EN	European Norm
MS	Mild Steel
CSI	Composite Sleeper Index
WBM	Water Bound Mcadam
mts/m	Metres

1

INTRODUCTION

Civil engineering is the branch of engineering that deals with design, construction and maintenance of structures, both natural and human made. Civil engineering was named this way so that it could separate itself from the branch of military engineering. Originally an engineer was one who was involved in making and operating an engine. Any engineering application related to civilian use was termed civil engineering and any engineering usage related to military was called Military engineering. In modern terms Civil engineering is the oldest branch of engineering.

In the early years of construction, the field of civil engineering and architecture were no different. The creator of structures had the knowledge of architecture, civil engineering, mechanics and astronomy. The designers ensured that the structure was constructed as per their vision and design. Hence the designers also had the practical knowledge of execution. They also had the knowledge of climatic conditions, movements of earth, the natural forces and changes which would affect the structure over a period of time. However as the society progressed and construction of structures rose, the field of architecture was separated from the field of civil engineering. Architects concentrated more on the creativity of shape and it was left to the engineers to convert that creativity into a reality.

1. HISTORY

Though the formal degrees of civil engineering came up in the late 18th to early 19th century, the concept is as old as civilizations when human began building shelters for various purposes such as the pyramids of Egypt, various townships, dams, waterways, lighthouses, forts, temples etc.

Till a few decades ago, there was no clear distinction between architecture and engineering. The engineers of the past often worked on technical problems of their day using mechanics and available materials. Technicians of the day worked in areas that ranged from mechanical, military inventions to architecture that involved the design of overall shape and form of structures which involved both aesthetics and practicality.

The Institution of civil engineers was formed in 1818 in London and it received the royal charter in 1828 thus recognizing it as a profession. The royal charter defined civil engineering as "the art of directing the great sources of power in nature for the use and convenience of man, as the means of production and of traffic in states, both for external and internal trade, as applied in the construction of roads, bridges, aqueducts, canals, river navigation and docks for internal intercourse and exchange, and in the construction of ports, harbours, moles, breakwaters and lighthouses, and in the art of navigation by artificial power for the purposes of commerce, and in the construction and application of machinery, and in the drainage of cities and towns".

The first private college to teach Civil Engineering in the United States was Norwich University founded in 1819 by Captain Alden Partridge. The first degree in Civil Engineering in the United States was awarded by Renssellaer Polytechnic Institute in 1835. The first such degree to be awarded to a woman was granted by Cornell University to Nora Stanton Blach in 1905.

In India first institution related to civil engineering was the school of survey in 1794 near Fort St. George, Madras (present day Chennai). This later became School of Civil Engineering in 1858 and further renamed as College of Engineering in 1859. The first formal engineering college to be established was Thomason College of civil engineering in Roorkee which is now known as Indian Institute of Technology, Roorkee. The first Indian to graduate from this college was Rai Bahadur Kanhaiya Lal. The first Indian woman civil engineer was Shakuntala A. Bhagat.

In Indian mythology the role of celestial architect or divine engineer is assigned to a god named Vishwakarma. He is said to have created cities and palaces that are eye-catching beauty even in today's world. It is said that the power of his vision is such that he would know exactly the life of the city or structure he built and how it would be destroyed. The creation of famed city of Dwaraka and Lanka is attributed to him. The festival commemorating Vishwakarma is celebrated on Kanya Sankranti which is the last day of the 5th month in Hindu calendar. It is believed that on this day, He invented the plough and gave it to humanity. The plough represents both the artisan trade as well as agriculture and therefore becomes the representative symbol of the ancient Indian civilization.

It would be a sin not to mention the name of M.Visveswaraya, one of the most prominent and renowned civil engineers of the country. His achievements include the construction of Krishanaraja sagar dam in Mysore and flood protection system in Hyderabad. To honour him, his birth anniversary, 15th September is celebrated as Engineers day.

The field of civil engineering itself started branching out to more specialized branches to concentrate more into the design and execution of complicated structures.

2. BRANCHES OF CIVIL ENGINEERING

Civil Engineering is further divided into branches based on specialization viz: Structural engineering; Geotechnical engineering; Construction engineering; Transportation engineering; Water-resources engineering; Surveying; and Environmental Engineering.

2.1 Structural Engineering

Structural engineering is a specialized branch of civil engineering which deals primarily with design of structures both RCC and steel. A structural engineer designs the strength of the structure. The role of structural engineer has become very important in today's scenario where the structures are going vertically up and horizontally more complicated. The field of structural engineering has evolved over the years with the further specialization branching into on shore, off shore structures, dynamics especially earthquakes.

There is no record of how far back the term of structural engineering came to be in use. In modern day, the field of structural engineering took shape during the industrial revolution. In the 1970's the field developed with the introduction of computer based applications for analysis and design.

2.2 Geotechnical Engineering

As the name suggest, it deals with the technical properties of the earth or more appropriately soil. Alternatively it is also called soil mechanics. This field is very important as one of the main components in design of structure is the bearing capacity of soil. The design of foundation is predominantly dependent on the safe bearing capacity (SBC) of soil.

Though humans have used soil for various activities in building foundations, flood control, irrigation etc., it was not until the 18th century that investigation of soil became more scientific and documented. The earliest advances in soil mechanics developed from the theories put forward by Henri Gautier and Charles Coulomb. The Mohr's-Coulomb theory on cohesion of soil is still in practice. In the 19th century, people like, Henry Darcy, Joseph Boussique and William Rankine proposed theories on earth pressure. Albert Atterberg developed indices for clay consistency, which is still used for classification of soil. Karl Von Terzaghi, an Austrian civil engineer of late nineteenth and early twentieth century is called the father of soil mechanics. His formula for determining the bearing capacity of soil is still valid today.

2.3 Construction Engineering

This branch of civil engineering deals with actual process of construction. This describes how the design is converted to reality for the structure to be put to use and also the do's and don'ts in the process. In a way we can say that this branch deals in making the dreams of architects and stakeholder come true. With structures becoming more complex and time being a criterion, new methods and materials are being evolved to facilitate the changes.

The introduction and development of project management concepts have taken the field of construction engineering a whole new dimension. Earlier, the process of documentation during the construction process was limited to measurements and billing. However, in today's scenario, documentation is a very important part of the process. The actual process is documented before the work is carried out, and evaluated for optimization. Every stage of work is documented. This also helps in post work evaluation of value engineering.

2.4 Transportation Engineering

This branch of civil engineering deals with providing the pathway for different modes of travel to ensure that the travel is comfortable, efficient, rapid, economical and safe. The field of transportation includes highway, railways, waterways and airways. Construction of roads, laying of rail tracks, building of canals and waterway and building of pavements for air and space travel, are a part of transportation. This field has great importance in urban planning and design. Design of cities and inter and intra city travel are complimentary.

Early human beings used animal for transportation over the land and raft like structures to travel on water. Boats were the earliest human made mode of transport. With the invention of wheel, carts and chariots became the mode of travel on land. It made travel much faster and less strenuous. The invention of engine and the technology to convert any form of energy to mechanical energy paved way for motorized vehicles as the mode of transport. The improved modes of transport needed proper pathways to ensure the travel to be smooth and comfortable. This facilitated the development of the transportation engineering as a field. The field of transportation has seen major improvements in the last three centuries.

2.5 Water-resources Engineering

Harnessing of natural resources has always been a continuous effort for humankind from the time civilizations started. This branch of civil engineering deals with largest natural resource namely water. Measuring from the simplest thing like the quantity of rainfall to greater things like building of water retaining structures like dams or canals is covered in this field. Water resources engineering is also known as hydrology.

The earliest use of hydrology was mainly for irrigation. The ancient hydrologist built massive canals with dams and levees to harness water for both irrigation as well as transportation. They used gravity to move water. In many aspects the principles used in modern hydrology remain the same. In most places, gravity method still remains the ways and means to move liquids. However the introduction of mechanical means has facilitated some difficult projects that have been highly beneficial to humankind. Panama Canal is one such example where the canal is higher than the sea level.

2.6 Surveying

This branch of civil engineering deals with measurements. The measurements range from measurement of the earth to measure of smaller units such as buildings or rooms or measuring for the purpose of estimation. Measuring for the purpose of estimation of work is generally termed as quantity surveying and that in itself is a specialization. The field of surveying has advanced over the years to adopt new instruments and technology which eliminate human error.

Surveying has been an element in the development of the human environment since the beginning of recorded history. The planning and execution of most forms of construction require it. It is also used in mapping, and the definition of legal boundaries for landownership.

The field of surveying has advanced by leaps and bounds with the introduction of GIS (Geographical Information System). Robert Tomlinson is considered as the father of GIS. GIS is a system designed to capture, store, manipulate, analyze, manage, and present spatial or geographical data.

2.7 Environmental Engineering

This branch of civil engineering developed much later and in more recent times when humankind industrialization started affecting the natural resources and the society started becoming aware of the need to protect the nature. The realization that the structures need to be in synergy with the nature and its surroundings for a long-term survival of humankind and prevention of total mechanization of lifestyle has made this field very prominent in today's scenario.

Development of human settlement always took into account the quality of environment surrounding their dwellings. History records that concept of sewers for waste disposal dates back to Harappan civilization. The Romans constructed aqueducts to ensure healthy water supply. The London sewer system built from 1849 to 1865 is another example of development in the field of environmental engineering. However it was only in 20th century that environmental engineering emerged as a separate discipline in response to the widespread concern on environmental degradation.

The field of environmental engineering deals with solid waste management, treatment of water supply and wastewater management, air pollution and effect on ecological balance. Today any major project needs have clearance from the local environmental agency to ensure that the development is proposed and carried out as per the set guidelines.

3. PROFESSIONS IN CIVIL ENGINEERING

The most common misconception is that civil engineers are involved only in the construction of buildings. However that is not true. The professions civil engineers are involved in wide and varied. Every activity has a specialized role and needs in-depth knowledge to carry out the activity. There is a lot of creativity involved in the projects civil engineers are involved in. The various professions the Civil Engineers are involved in are as follows:

3.1 Design Engineer

A design engineer primarily deals in finding out the dimension of a structural member and amount and type of materials which will be sufficient to withstand different types of loads and forces applied on that member, at the same time it will be economic and providing serviceability. The design process is an information intensive one and also involves a good amount of analytical work combined with the practicality of execution. In civil engineering the design engineer is generally known as structural design engineer or structural engineer. A structural engineer should have keen interest in analytical work. A designer's role is generally in office.

In designs of building, the structural engineer works closely with the architect. An architect designs the layout of the building or structure and the structural engineer designs the framework. Architect and the structural engineer have to complement each other to ensure that the final design of structure satisfies both the strength as well as aesthetics. However in structures like towers and bridges, where strength of the structure is the main criteria, it is the structural engineer who takes precedence.

Structural design involves two main categories, viz. design of steel structures and design of concrete structures. Though the fundamental principle of analysis remains the same for both concrete as well as steel, the process of design is different due to material properties and their behavior under stress.

3.2 Geotechnical Engineer

A geotechnical engineer role is to investigate and assess the sub-surface and further analysis to determine its load carrying capacity for various works. Geotechnical engineer's roles also involve in evaluating the stability of slopes and assess the risk posed by site conditions during earthwork. Geotechnical engineer is also known as soil mechanics engineer or in short soil engineer.

Geotechnical engineer's role doesn't limit itself just to analysis of property of soil and suggesting foundation for the buildings. They role extends to irrigation and environmental issues too. The determination of stability of slope is an important factor in the design of canals and embankments.

The role of geotechnical engineer involves both site and office work. The process of collecting the samples and assessing the site condition is done at site and the analysis and design is done at office or laboratory.

3.3 Land Surveyor

Surveyor is a person whose occupation is determining the size, shape and boundaries of pieces of land. Surveyor is also known as land surveyor. A land surveyor's job is to measure and produce plans of an area. The area may be a small piece of land where one needs to build a house or a large piece of land where a township needs to be developed.

Land surveyors work with engineers, architects and builders to produce precise descriptions (surveys and maps) of surface features of the Earth. Land surveyors perform a variety of vital tasks such as boundary surveys, topographic mapping and construction staking.

Surveying is an onsite job where the surveyors are involved in measuring and collecting data. The office work is limited to preparation of final drawings and in modern days feeding data to computer.

3.4 Quantity Surveyor

Quantity surveyor (QS) is a person who calculates the amount of materials needed for building work, and how much they will cost. This field is different from land surveying. They are also called as estimators or cost engineers. They are also known as billing engineers in some cases. Quantity Surveyor is supposed to be an expert in costs and contracts. The role of QS is very important in any organization or project.

This role includes cost estimating and planning, budgeting, commercial management and contract administration and asset capitalization to name a few. QS must have sound knowledge of mensuration, trigonometry, and building construction. They must also be able to describe the specifications of each item clearly. This role is office based and QS visits site only to verify the measurements.

3.5 Valuer

A valuer is a person who estimates the value of an asset. His job is to say how much something is worth. In civil engineering a valuer estimates the value of immovable properties namely land and a building. A valuer should have the knowledge of quantity surveying.

The role of valuer in civil engineering is limited to valuation of immovable properties (land and buildings). In India the authorized body to register as a valuer is "The Institution of valuers".

In the ancient times, the value of a property was estimated based on the need of the parties involved in transaction. The value of a holding property was determined ad hoc rather than following a prescribed system. With the introduction of organized banking system and the double entry system of book keeping, it became necessary to follow a regularized valuation procedure. Every bank has a panel of valuers. Valuers also help in estimating a cost of property for loans and also for insurance.

A valuer has to visit site to ascertain the present condition and collect relevant data. After that it is mainly an office based operation where calculations and estimations are done.

3.6 Planning Engineer

Planning engineer, also known as project planner is responsible for planning the schedule or timelines of the project. Planning engineer, along with QS is a part of the project control team and it is a key area in project management. A planning engineer prepares the initial schedule of the project and then monitors the project on its progress by collecting the real time data from site and ascertaining the current position with respect to the timelines planned.

Planning engineers' role is office based. Their scope of role vary from preparing project schedules, generate executive reports on cash flow, keep a track on the progress to analyse time and cost overrun. They get their input data from QS and the site team. Planning engineers should have keen interest in collating and analyzing data.

3.7 Construction Engineer

A construction engineer is involved in managing and overseeing a civil engineering project. Construction engineers are a key cog in the field of civil engineering. They are responsible for converting dreams into reality. They should have the ability to read and understand drawings, manage people and time and also have good knowledge of quality and public health and safety. Construction engineers should also have good people management skills as he has to manage teams on site.

A role of construction engineer is predominantly site based. Office work is limited to recording and preparation of daily reports. Their inputs and feedback on the practicality of execution and real time progress is very important in project management and value engineering.

3.8 Quality Control Engineer

QC Engineers (Quality control engineers) ensure that the materials and process used in construction are adhering to the established and accepted standards. Sometimes they are also referred to as QA/QC engineers. The role of QC engineer has two parts, in site as well as in laboratory.

QC engineers, in a construction setup, work alongside the operations or execution team and help the team look specifically into quality allowing the execution team to concentrate on the progress of the project. They also act as an audit team to maintain quality in all aspects at every stage.

QC engineer's role in laboratory set up is to ensure the quality of material. Engineer in charge of the lab should be well versed in the standards of material specifications and the testing process. For larger construction, a lab is set up on site to carry out certain test procedures.

3.9 Transportation Engineer

Transportation engineer is involved in planning, design, construction, maintenance, and operation of transportation facilities. These facilities include air, highway, railroad, water, and even space transportation.

Transportation engineers are of paramount importance when it comes to urban planning and design. Modern day urban planning is depended on creating a seamless network for mobility. Frequency of trips, trip distribution, mode of travel and route assignments are planned and calculated by transportation engineers.

The design aspects of transport engineering include the sizing of transportation facilities, determining the materials used, designing the shape of path and determining the load carrying capacity.

3.10 Highway Engineer

Highway engineering is a specialized part of transportation engineering which deals only with roads. Highway engineers are involved in the planning, design, construction, operation and maintenance of roads and bridges and tunnels to ensure safe and effective transportation of people and goods.

The planning involves the selection of appropriate location for highways, alignment and shape. The design aspects of transport engineering include the sizing of transportation facilities (how many lanes or how much capacity the facility has), determining the materials and thickness used in pavement and designing the geometry (vertical and horizontal alignment).

3.11 Lab Technician

Engineering lab technicians help scientists and engineers to assess and solve technical problems related to materials. Every specialization of engineering requires a dedicated lab and the engineers handling it should have a good knowledge of materials and their properties. They should be thorough with the standards and codes of the region in which they are working. They should also have a keen interest in research.

With the increasing importance on quality compliances, the role of lab technicians has enhanced. All the major construction projects have a site laboratory to conduct preliminary testing on material to be used.

3.12 Drafter

Drawing is the language of engineers. Drafter or drafting technician is a person who makes detailed drawings of plans and sections of buildings, structures. In the past, the drafters prepared the drawings manually using equipments like, scale, protractor, triangles etc. With the advent of technology, the drafters used computers to draw and store drawings. The most popular software used for drafting is CAD (computer aided design) and drafters are also sometimes referred to as CAD operators.

Drafting is an office based work. It is important for a drafter to understand what the designer wants to represent in the drawing. Drafting requires high levels of concentration and patience. Any mistakes made by the drafter will result in incorrect execution.

4. ENGINEERS AND SOCIETY

Engineering being a field of application has its obligation to the society. Engineers design and construct structures which represent the prevalent culture, need and socioeconomic stature. The greatest merit of an engineer is the recognition from the society on the professional and commitment to the projects which take care of welfare of the people. Even Nobel Prize is conferred on discoveries which have greatest benefit to society.

Engineers have an important role in mediating between the values of society, clients, the environment and the possibilities of technology. Understanding engineering as a hybrid, socio-technical profession can help develop new ways of working that acknowledge the importance of technology and infrastructure in shaping social relationships that are central to achieving sustainability.

4.1 Engineering Bodies

The rapid rate of industrialization and the spiraling growth of engineering as a profession in the 19th century created a need to regularize the profession and practices. Also there was a need to differentiate between scientists and engineers. In 1771, led by John Smeaton, a group of people calling themselves as civil engineers formed the society of civil engineers. This laid the foundation for the idea of forming of the first professional engineering body called "Institution of Civil Engineers". In 1818, three young engineers, Henry Robinson Parmer, James Jones and Joshua Field took initiative to establish the institution of engineers. However it was only in 1828 and with the initiative of Thomas Telford, the first president of the institution that it received the royal charter and got its official recognition.

In India, the Institution of engineers was established on September 13, 1920 in Calcutta (now Kolkata) with Sir Thomas R.J. Ward as its first president. It received its royal charter in 1935. In 1921, Sir R.N. Mookherjee was elected as the inaugural president through an AGM.

As with other professions, the professional status and actual practice of engineering is legalized. This is done by permitting licensed or registered engineer to take legal responsibility for engineering works. Regulations may require that only a licensed or registered engineer can sign, seal, or stamp technical documentation such as reports, plans, engineering drawings, and calculations for study estimate or valuation, or carry out design, analysis, repair, servicing, maintenance, or supervision of engineering work, process or project.

4.2 Engineering Ethics

Engineering Ethics is defined as the moral principle that governs a person's behavior in the practice of engineering profession. The Code of Ethics is based on broad principles of truth, honesty, justice, trustworthiness, competence and accountability. It also includes respect for human values and safeguard and welfare of human life. Every Corporate Member of an engineering Institution must recognize, uphold and abide by the moral values of the engineering profession. There are several ethical issues an engineer will face. These include:

- Relationships with clients, consultants, contractors, competitors.
- Conflict of interest or Moonlighting.
- Ensuring legal compliance by clients, client's contractors, and others.

The broad principles of code of conduct for all engineers are as below:

- Engineers shall hold paramount the safety, health and welfare of the public and shall strive to comply with the principles of sustainable development in the performance of their professional duties.
- Engineers shall perform services only in areas of their competence.
- Engineers shall issue public statements only in an objective and truthful manner.
- Engineers shall act in professional matters for each employer or client as faithful agents or trustees, and shall avoid conflicts of interest.
- Engineers shall build their professional reputation on the merit of their services and shall not compete unfairly with others.
- Engineers shall act in such a manner as to uphold and enhance the honor, integrity, and dignity of the engineering profession and shall act with zero-tolerance for bribery, fraud, and corruption.

- Engineers shall continue their professional development throughout their careers, and shall provide opportunities for the professional development of those engineers under their supervision.

Reference and Further Reads

Jayshree, Suresh, Raghavan B.S., "Human Values and Professional Ethics", 2007, S. Chand & Company Ltd.

Govindarajan, M., S. Natarajan, V.S. Senthilkumar, "Engineering Ethics", 2004, Prentice Hall International.

Okumu, Valdengrave, "An Introduction to Civil Engineering", 2014, Createspace Independent Publishing Platform.

Wood, David Muir, "Civil Engineering: A very short introduction", 2012, Oxford University Press.

2

BUILDING MATERIALS

There are various types of materials used in construction for both structural use and for finishing. Some materials are used in combination with other materials to manufacture a third material. In earlier days, due to the availability of time and resources, the materials available in nature were used for construction directly. As the number of construction increased both in size and complexity of shapes, the natural materials were found inadequate to serve the purpose. Subsequently new materials were manufactured by processing the raw material either individually of combining with other raw materials to compensate for the drawback of one material and ensuring that the new material has the good qualities of both the materials.

1. BUILDING BLOCKS

Building blocks are materials used in masonry construction and are usually laid using different type of mortars for binding. The history of using bricks dates back to 7500 BC, when brick and stones were the primary construction material with lime mortar or gypsum mortar as binding material. Materials like stone and timber were used in their natural state. The advent of industrialization paved way for manufacture of artificial materials. The manufacture of cement and introduction of concrete in construction made concrete blocks and their variance as a viable option of building material.

1.1 Stone

Stones were the earliest building materials to be used in construction. Stones are present in their natural state and are abundantly available and hence it was easy to procure. The biggest advantage stones had was the durability to withstand the weathering caused by natural forces. Most of the structures built using stones in earlier days have withstood the test of time.

Stones also geologically known as rocks have three basic classification namely Igneous, sedimentary and metamorphic. The formation of stones or rocks is a very long process and it commenced when the gases present in the earth began to liquefy and then subsequently solidify.

(a) **Igneous Rocks:** Igneous rocks are formed due to the compression of heat of the earth's crust. These stones originate from the crystallization of molten rocks. The proportion of crystal determines the colour and characteristics of the stones. Most of the stones used for construction especially for structural use are igneous rocks. Granite is a form of igneous rock and most commonly available. Granites are mainly composed of quartz, feldspar and mica. The biggest advantage of granite is its strength and hardness. It can withstand the abrasion of weathering. In earlier days almost all mega structures were built using granite.

(b) **Sedimentary Rocks:** Sedimentary rocks are a softer variety of rock. They are the type of rocks that are formed by the accumulation of organic particles either on land or in water. Typical examples for sedimentary rocks are sandstone, limestone and shale. Sandstones are stratified rocks produced from disintegrated igneous rocks. These types of rocks are generally used for cladding or flooring where there is a harder base to support it.

(c) **Metamorphic Rocks:** Metamorphic rocks are rocks formed due to the transformation of either igneous or sedimentary rocks under heavy pressure and temperature. The process of change is known as metamorphism and hence the name metamorphic. Marble and Slate are a type of metamorphic rock transformed out of shale which is a sedimentary rock. Marbles are produced from limestone. Similar to sedimentary rocks, metamorphic rocks are used as flooring or cladding materials. However in olden days marble blocks have been used as structural elements. Taj Mahal is built using marble blocks.

1.2 Clay Bricks

Bricks were the first human made material to replace stones as a construction material. Bricks can be handmade or machine moulded. The modern clay bricks are made of silica (sand), alumina (clay), lime, iron-oxide and manganese. The elements are mixed and pressed into moulds using a hydraulic press and fired to a temperature of about 900-1000°C.

In India IS1077 gives the description of common burnt clay building bricks. The standard size of a clay brick is 230 × 115 × 75 mm (9 × 4.5 × 3 inches).

A brick of good quality should:

- Be uniform in colour and texture
- Have smooth rectangular faces with sharp corners
- Emit a clear ringing sound when stuck
- Have good compressive strength
- Have higher bulk density
- Have water absorption less than 10% by weight

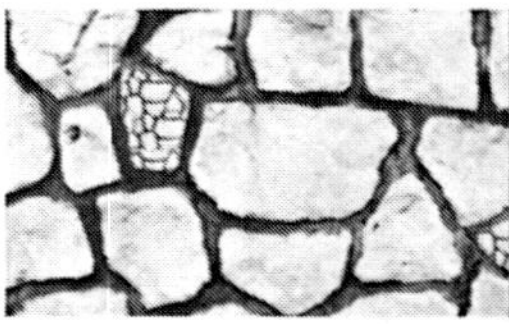

Fig. 2.1(a): Stone **Fig. 2.1(b):** Clay bricks

Fig. 2.1(c): Cement concrte blocks **Fig. 2.1(d):** AAC blocks

Fig. 2.1(e): Adobe bricks **Fig. 2.1(f):** Flyash bricks

Fig. 2.1: Figure showing building blocks

1.3 Cement Concrete Blocks

The cement concrete blocks have replaced clay bricks in most of the projects. The lack of availability of good quality clay and increasing prices of clay bricks has facilitated this change. Cement concrete blocks are made of concrete. The materials used are similar to concrete but the only difference is the course aggregates used is of size of 12 mm or less. Concrete blocks are often made of 1:3:6 (cement : sand : coarse aggregates) concrete with a maximum size aggregate of 10 mm or a cement-aggregate mixture with a ratio of 1:7, 1:8 or 1:9. Under no circumstances the mix should be richer than 1:6 by volume. The size of a standard block is 400 mm × 200 mm (length × height) and the thickness varies from 100 mm, 150 mm and 200 mm. The specifications for concrete masonry unit are described in IS2185.

The cement concrete blocks are of two types namely: Solid concrete blocks and Hollow concrete blocks. The hollow concrete blocks is a very popular material in the present day construction of non-load bearing wall like compound wall. Generally solid concrete block of thickness higher than 150 mm are avoided due the problems it poses in handling. However 200 mm thick solid blocks are used for raising of plinth (basement as it is popularly called in India) where the wall size in superstructure is 200 mm. The advantage of concrete block is its low cost and higher strength.

1.4 AAC Blocks

AAC Blocks (Autoclaved Aerated Concrete blocks) are lightweight concrete blocks that are manufactured by injecting vacuum in concrete blocks during manufacturing. In this process the voids are replaced by vacuum/foam, thus reducing the self-weight of the block. The standard size of an AAC block is 600 mm × 200 mm (length × height). The thickness varies from 100 mm, 150 mm, 200 mm and 230 mm. The dry density of the block is 6-8 kN/m^3. Due to its light weight, the overall cost of structure can be reduced by 10% compared to the use of bricks. AAC blocks are becoming the most popular material to be used for masonry especially in multi-storeyed buildings. The advantages of the block are as follows:

- The blocks are lightweight and hence reducing the total weight of the structure.
- Since the blocks are large sized, the numbers of joints are less and hence less consumption of cement mortar.
- Being lightweight, handling is easier and construction is faster.
- The finish of the blocks being uniform and smooth, the plastering is uniform and consumes less mortar.
- Due to the presence of vacuum, it has very good acoustics.
- It has very good fire resistive properties.

1.5 Adobe Bricks

Adobe bricks are building blocks made out of earth and sometimes organic material. Adobe in Spanish means mud bricks. The appearance of these bricks is similar to rammed earth. Adobe is generally made by mixing locally available earth and some organic material like straw or dung. The history of adobe blocks dates back to thousands of years across the world. Composition of adobe consists of 15 – 25% clay, 10 – 30% silt and 55 – 75% sand. Modern adobe is stabilized either by mixing asphalt or Portland cement.

Adobe is an inexpensive material with very little resource cost. Due to its thermal properties, it is a good material for construction in desert climates. In dry climates adobe has very good durability. Care should be taken to ensure that the clay used does not have expansive properties. They are also susceptible to earthquakes.

1.6 Fly Ash Bricks

Fly ash brick is a building material made primarily of fly ash. It is a relatively new material with the production history dating back to the 19th century when coal was widely used as a form of energy. The mixture of fly ash, sand or stone dust, OPC/gypsum and water is compressed at 28 MPa and cured for 24 hours in a 66°C steam bath. Owing to the high concentration of calcium oxide in class C fly ash, the brick is sometimes described as "self-cementing". The manufacturing method saves energy, reduces mercury pollution, and costs 20% less than traditional clay brick manufacturing.

Composition of flyash bricks consists of 60 – 80% fly ash, 10 – 20% lime, 10% of gypsum and 10% sand depending on the quality of raw material. The advantages of these bricks are:

- Light weight and high strength
- Less mortar consumption due to uniform size
- Good fire insulation properties.

However it has low mechanical strength and there is a limitation on the size of the blocks.

2. WOOD/TIMBER

After stones, wood was the next most popular natural material used for construction. Wood was abundantly available as the area covered by forest was vast. Unlike stones, which could be used in its natural state, wood needs to be processed into planks and beam which is called timber. Wood is used for various activities in construction starting from formwork to construction of buildings. Prior to 19th century (in the medieval age) wood was most widely used material for construction of houses. Timber is good for construction for a single to two storied house. The foundation is generally made of stone and the superstructure is made of timber. Wood has also been used as foundations namely pile foundation.

Some of the woods used for different types of construction are described in Table 2.1.

Table 2.1: Table showing the types of wood and their uses

S. No.	Wood	Usage
1	Teak	Structures, Joineries, Furnitures
2	Padak	Joineries
3	Sal/Kongu	Joineries
4	Neem	Joineries
5	Rose wood	Joineries, Furnitures
6	Marandi	Joineries
7	Casuarina	Scaffolding, Formwork Support
8	Silver wood	Formwork
9	Karungali (tamil)	Structure, Joineries
10	Vengai (tamil)	Joineries

2.1 Benefit of Timber

Timber is the most ecofriendly material that can be used for construction. Even though it is expensive in present day scenario, its benefits are multi-fold. The benefits of using timber are:

- It requires less energy to convert from raw material to finished product.
- It degrades naturally and does not cause any harmful effect to the environment.
- It is a good reservoir of atmospheric carbon.
- It is a good insulator and reduces the energy required to heat or cool the inside of the building.
- Construction is faster and lighter.
- The temperature affects are minimal when locally available timber is used which is already seasoned.

2.2 Properties of Good Timber

A good timber should:

- Have a uniform colour.
- Have a pleasant odour or smell when cut freshly.
- Give a clear ringing sound when struck.
- Be of fine and even texture.
- Have closer grain structure.
- Have higher density for better strength.
- Be harder timbers for better strength and durability.
- Not warp under changing environmental conditions.
- Have higher toughness to withstand shock loads.
- Not deteriorate due to wear. This property should be kept in mind when used for flooring.
- Have high strength in bending, shear and direct compression.
- Have high modulus of elasticity especially when used for construction activities.
- Have high resistance to fire.
- Have low water permeability.
- Be easily workable. It should not clog the saw.
- Resist the action of fungi and insects attack.
- Be free from defects like dead knots, shakes and crack.

2.3 Classification of Timber

Various bases are considered for the classification of timbers. The following are the important basis:

(a) Mode of growth

(b) Modulus of elasticity

(c) Durability

(d) Grading

(e) Availability.

(a) **Classification Based on Mode of Growth:** On the basis of mode of growth trees are classified as

(i) **Exogeneous Trees:** These trees grow by the addition of new wood outwards, immediately under the bark.

(ii) **Endogenous Trees:** Endogenous trees grow by forming new fibers within the trunk interspersed with the old fibers.

(b) **Classification Based on Strength of Wood:** Based on the strength of wood, it can be classified as soft wood and hard wood. The classification as soft wood and hard wood has more of a commercial importance. The difference between soft wood and hard wood is shown in Table 2.2:

Table 2.2: Table showing the difference between hard and soft wood

Property	Soft wood	Hard wood
Rings	Distinctly visible	Indistinct
Colour	Lighter	Darker
Compressive Strength	Less	More
Shear Strength	Less	More
Structure	Resinous	Close grained
Fire Resistance	Poor	Good

(c) **Classification Based on Modulus of Elasticity:** Elastic modulus is the object's resistance to deformation when a force is applied on it. Young's modulus is determined by conducting bending test. On this basis timber is classified as:

- Group A: $E = 12.5$ kN/mm^2
- Group B: $E = 9.8$ kN/mm^2 to 12.5 kN/mm^2
- Group C: $E = 5.6$ kN/mm^2 to 9.8 kN/mm^2.

(d) **Classification Based on Durability:** Durability of timber is defined as the ability to resist decay due to nature and attack from pests. Durability tests are conducted by the forest research establishment. In this test the specimens of size 600 × 50 × 50 mm is buried in the ground to half their length and their conditions are observed over a period of time. The timbers are classified as:

- High durability: If average life is more than 10 years.
- Moderate durability: If average life is between 5 to 10 years.
- Low durability: If average life is less than 5 years.

(e) **Classification Based on Grading:** Structural grading is the process by which timber is sorted into groups with ideally similar structural properties in each group. The classification is based on permissible stresses, defects etc. Grading into different strength classes allows the best possible use to be made of the various timber dimensions. This ensures that timber that is unsuitable due to low strength, twist, warp or springs is not used structurally. IS883 classifies the structural timber into three grades namely

- Select grade
- Grade I
- Grade II.

The various other requirements of structural timber should conform to IS3629.

(f) **Classification Based on Availability:** Forest departments classify timbers based on the availability as

- X—Most common. 1415 m^3 or more per year
- Y—Common. 355 m^3 to 1415 m^3 per year
- Z—Less common. Less than 355 m^3 per year.

3. CONCRETE

Concrete is an artificially manufactured material which has strength almost equal to a stone. It is manufactured by combining cement, fine and coarse aggregates and water in a predetermined ratio. In some cases additives, known as admixtures, are added to enhance certain special properties. The ratio of mix is specified in IS456.

Concrete has very good compressive strength but very low tensile strength. The tensile strength is imparted to the concrete either by reinforcing it with a very good tensile material name steel or the tensile stresses is induced internally. The former is known as Reinforced Cement Concrete (RCC) and the latter is known as Pre-stressed concrete.

The properties of concrete are dependent on the individual properties of the ingredients and also the ratio in which they are mixed. The ingredients of cement concrete are Cement, Aggregates, Water, and Admixtures.

3.1 Cement

Cement is a binding material used to construct buildings where constructional units are bonded together with the help of cement mortar. Cement can be described as the building material with adhesive and cohesive properties which make it capable of bonding material fragments into solid mass of adequate strength and durability.

Cement was first manufactured and patented by Joseph Aspdin on 21st October 1824 in England. The product was called Portland cement because the set mortar made from it resembled the Portland stone. Portland stone was the most commonly used construction material used in England during that time. Before 1824, lime was used as binding material. All the structures were built using lime mortar.

Fig. 2.2(a): Cement bags

Fig. 2.2(b): Course aggregate

Fig. 2.2(c): Fine aggregate

Fig. 2.2: Figure showing raw materials of concrete

Portland cement is produced by burning together a mixture of naturally occurring argillaceous and calcareous at temperature range of 1300 to 1500°C. The product thus obtained is clinker. The clinker is the cooled and a small quantity of gypsum is added. The mixture is ground to a form of fine powder which is cement.

The calcareous portion consists of lime, chalk, marl and marine shell.

Argillaceous portion consists of clay, shell and blast furnace slag.

3.1.1 Methods of Manufacturing Cement

There are two methods to manufacture cement.

(a) **The dry method:** In this method the constituents of cement, i.e. the calcareous and argillaceous materials are mixed in desired proportion and fed into the blast furnace or kiln where the mixture is heated to a very high temperature of about 1400°C. The product thus obtained is clinker which is then cooled and pulverized or crushed into powder. The powder obtained is cement.

(b) **The wet method:** In this method the limestone is broken into pieces with the help of crusher and then stored in a basin. Simultaneously clay is mixed with water to form slurry. The broken pieces of limestone and clay slurry are mixed in a grinder in a predetermined ratio.

3.1.2 Types of Cement

Cement is of the following types:

(a) Portland cement
(b) High alumina cement
(c) Super-sulphate cement
(d) Natural cement
(e) Special cement

(a) **Portland Cement:** Following are the types of Portland cement

(i) Ordinary Portland cement
(ii) Rapid hardening Portland cement
(iii) Low heat Portland cement
(iv) Portland blast furnace slag cement
(v) Portland pozzolonic cement
(vi) Sulphate resisting Portland cement
(vii) White Portland cement
(viii) Coloured Portland cement.

(i) **Ordinary Portland Cement (OPC):** It is the most common type of cement used in construction. This is the basic form of cement and other variants of cement are manufactured either by changing the composition of OPC or by adding some extra chemicals. The percentage of ingredients and compound of clinker and properties are described in Table 2.3.

The ordinary Portland cement has 3 grades—33, 43 and 53, the numbers denoting the compressive strength at 28 days. The properties of OPC have been specified in various IS codes.

- IS269 for 33 grade cement
- IS8112 for 43 grade cement
- IS12269 for 53 grade cement

Table 2.3: Table showing the ingredients of Ordinary Portland Cement

Ingredients		Clinker		Properties		
Mineral	%	Clinker compound	%	Property		Unit
Lime	63.0	Tricalcium silicate	40	Specific surface	2250	cm^2/gm
Silica	20.6	Dicalcium silicate	30	Initial setting time	30	mins
Alumina	6.3	Tricalcium Aluminate	11	Final setting time	600	mins
Iron Oxide	3.6	Tetracalcium Alumino ferrite	11	Ave 28-day compressive strength	16-22	MPa

(ii) **Rapid Hardening Portland Cement:** As the name suggests, this type of cement sets faster than the normal Portland cement. Generally the strength developed in 3 days for this type of cement is equal to the 7-day strength of OPC. The advantage of using this is cement is that the formwork can be removed earlier and structure could be put to use earlier. The percentage of ingredients and compound of clinker and properties are shown in Table 2.4.

Table 2.4: Table showing the ingredients of Rapid Hardening Portland Cement

Ingredients		Clinker		Properties		
Mineral	%	Clinker compound	%	Property		Unit
Lime	64.5	Tricalcium silicate	50	Specific surface	3250	cm^2/gm
Silica	20.8	Dicalcium silicate	21	Initial setting time	30	mins
Alumina	5.2	Tricalcium Aluminate	9	Final setting time	600	mins
Iron Oxide	2.9	Tetracalcium Alumino ferrite	9	Ave 28-day compressive strength	16-22	MPa

If the setting time in this type of cement has to be further reduced, then the mixture is added with calcium chloride. An addition of 2% of $CaCl_2$ reduces the initial setting time by 10 minutes and increases the strength by 25%.

(iii) **Low Heat Portland Cement:** This type of cement is less reactive than the OPC and gives out lesser heat of hydration. This type of cement develops strength at a slower pace than OPC. The rate of evolution of heat for 7 days is 65 cal/gm and for 28 days is 75 cal/gm. It is generally used where mass concreting is done especially gravity dams. The percentage of ingredients and compound of clinker and properties are shown in Table 2.5.

Table 2.5: Table showing the ingredients of low heat Portland Cement

Ingredients		Clinker		Properties		
Mineral	%	Clinker compound	%	Property		Unit
Lime	60.5	Tricalcium silicate	25	Specific surface	2250	cm^2/gm
Silica	22.5	Dicalcium silicate	45	Initial setting time	30	mins
Alumina	5.2	Tricalcium Aluminate	5	Final setting time	600	mins
Iron Oxide	4.6	Tetracalcium Alumino ferrite	14	Ave 28-day compressive strength	16-22	MPa

(iv) **Portland Blast Furnace Slag Cement:** In this type of cement ground blast furnace slag is added to OPC. IS455 describes the details and specifications of this type of cement. This type of cement is similar to low heat Portland cement. Due to the replacement of cement clickers with GGBS, the cost of this cement is lower than OPC. It is generally used where mass concreting is done especially gravity dams. This type of cement should not be used in cold weather concreting.

(v) **Portland Pozzolona Cement:** Portland Pozzolona Cement or PPC, as is commonly known as, is manufactured by addition of pozzolona to the cement clinkers. IS1489 describes the details and specifications of this type of cement. Pozzolona is nothing but fly ash or calcinated clay. This type of cement develops less heat of hydration and also has higher resistance to reactive chemical agents. It is generally used in mass concreting of hydraulic structures prone to chemical attack. PPC takes a longer time to set and hence it is best to avoid this type of cement where removal of formwork is to be carried out early. Generally PPC doesn't have high initial strength and gains strength after 28 days. This type of cement should not be used in cold weather concreting.

(vi) **Sulphate Resisting Portland Cement:** This type of cement is manufactured by grinding Portland cement clinkers with gypsum and furnace slag. In this type of cement the percentage of C_3A and C_4AF is comparatively low and has high silicate content. This cement is used in places where sulphate attack is expected especially marine structures and sewage treatment plants. The percentage of ingredients and compound of clinker and properties are shown in Table 2.6.

Table 2.6: Table showing the ingredients of Sulphate resisting Portland Cement

Ingredients		Clinker		Properties		
Mineral	%	Clinker compound	%	Property		Unit
Lime	64.0	Tricalcium silicate	40	Specific surface	4000	cm^2/gm
Silica	24.5	Dicalcium silicate	40	Initial setting time	30	mins
Alumina	3.7	Tricalcium Aluminate	5	Final setting time	600	mins
Iron Oxide	3.0	Tetracalcium Alumino ferrite	9	Ave 28-day compressive strength	15-30	MPa

(vii) **White Portland Cement:** It is similar to OPC in all aspects except that china clay (white clay) is added to the constituents. The quantity of iron oxide is limited to 1%. The properties of white portland cement have been specified in IS8042. Generally it is not used for structural works. It is used for plastering and ornamental works. It is also used in Terrazo flooring.

(viii) **Coloured Portland Cement:** It is similar to white cement, except that colour pigments are added to cement clinker. The percentage of pigment which should be added should not exceed 10%. They are widely used in terrazzo flooring finishes, pointing of marbles or tiles, external surfaces, textured panel surfaces etc. Table 2.7 describes the chemicals added for various colours.

Table 2.7: Table showing the colouring agents of Coloured Portland Cement

Colour	Mineral	Ratio (kg/bag)
Green	Chromium Oxide	2.7
Red	Turkey red	2.3 to 4.1
Brick (brownish red)	Iron Oxide	2.3 to 4.1
Black	Manganese Oxide	0.5 to 1.0

(b) **High Alumina Cement (HAC):** As the name suggest, this type of cement has high alumina content. This type of cement is highly reactive and produces very high early strength. It is manufactured by mixing bauxite, lime, iron oxide, and silica and ground to fine powder. The fineness of this cement is higher than that of OPC. It can gain 80% of its strength in 24 hrs. Because of its quick setting properties, the formwork can be removed early. However due to the high cost of bauxite, this cement is expensive.

HAC is characterized by it dark colour, high early strength, high heat of hydration and good resistance to chemical attacks. The properties are specified in IS6452. The percentage of ingredients and compound of clinker and properties are described in Table 2.8.

Table 2.8: Table showing the ingredients of High Alumina Cement

Ingredients		**Clinker**		**Properties**		
Mineral	%	Clinker compound	%	Property		Unit
Lime	38.0	Tricalcium silicate	40	Specific surface	2500	cm^2/gm
Silica	6.0	Dicalcium silicate	40	Initial setting time	30	mins
Alumina	30.0	Tricalcium Aluminate	5	Final setting time	300	mins
Iron Oxide	10.0	Tetracalcium Alumino ferrite	9	Ave 28-day compressive strength	15-30	MPa

(c) **Super Sulphate Cement:** This type of cement is made of blast furnace slag (80 – 85%), calcium sulphate (10 – 15%) and OPC (1 – 2%). Its fineness is higher than OPC. Its most important property is its low heat of hydration, hence it is suitable for works which have mass concreting like dams. The expansion and contraction properties are volatile and hence care should be taken during curing. It also has good resistance to chemical attacks. The properties of this cement are specified in IS6909.

(d) **Natural Cement:** Natural cement is that type of cement which is manufactured from a naturally occurring limestone. These stones inherently have the property of calcarious and argilacious materials. One of the examples of natural cement is Rosendale cement. Because of its low heat of hydration, it is a good cement to be used in mass concreting.

(e) **Special Cement:** Special cement is that type of cement which is manufactured specifically to resist properties of special uses. The types of special cements are as follows:

   (i) **Oil well Cement:** Generally the process of boring for oil well done in rocks. Boring in rocks can result in gaps between the steel casing and the rock. The presence of these gaps can allow the gas or oil to seep out. To plug these gaps oil well cement is used. Oil well cement is a type of hydraulic cement which has a slow setting rate under high temperature.

   Oil well cement is manufactured by adjusting the proportion of C_3A and adding retarders. Reducing the C_3A content can help in increasing the setting time.

   (ii) **Masonry Cement:** In earlier days, lime mortar was used in masonry as a binder. Lime mortar has a very good workability. To give similar properties to lime mortar and also enhance the durability, masonry cement is manufactured. In this the ratio of cement, lime and sand is 1:1:6. The advantage of this type of cement is its increased durability and low shrinkage as compared to OPC.

   (iii) **Expansive Cement:** Expansive cement is that type of cement which expands during hardening. Generally cement concrete has shrinkage properties which results in shrinkage cracks. To compensate this, the cement is induced with expansive properties which will compensate the drying shrinkage. This type of cement exhibits good expansion properties in presence of moisture too. This type of cement can be used in roof slabs, canal linings and in waterproofing and repair of concrete.

   Expansive cement is manufactured by mixing sulpho-aluminate to OPC. The percentage of sulpho-aluminate mixed to OPC is in the range of 8 to 20% and the total expansion ranges between 0.2 to 2%.

3.2 Aggregates

Aggregates are chemically inert materials which, when bonded with cement paste form cement concrete. Aggregates constitute to the strength of the concrete because they constitute the bulk of volume of concrete.

Following are the functions of aggregates:

- It helps cement concrete to behave like a artificial stone.
- It provides bulk volume and greater volume stability.
- It provides durability to concrete.
- It increases the density of concrete.
- It increases workability of concrete.
- It prevents segregation.

Following are the requirements of aggregates which are to be used for construction purposes:

- It must be of required shape and size.
- It must be clean.
- It must be hard.
- It must be strong.
- It must be well graded.
- It must be chemically inert.

3.2.1 Classification of Aggregates

Aggregates are classified into four main categories:

(a) Classification based on geological origin.
(b) Classification based on shape of aggregates.
(c) Classification based on size of aggregates.
(d) Classification based on unit weight of aggregates.

(a) **Classification Based on Geological Origin:** Aggregates can be divided into two categories based on the origin of the sources namely Natural aggregates and Artificial aggregates:

  (i) **Natural Aggregates:** As the name suggests, the natural aggregates are those which are found in nature. The examples of natural aggregates are sand, gravel, crushed rock like granite, basalt, sandstone etc. The specifications of natural aggregates are specified in IS383—1970. As mentioned earlier, in geology there are three types of rocks namely Sedimentary, Igneous and Metamorphic.

  (ii) **Artificial Aggregates:** In certain instances, it would be difficult to procure natural aggregates. Under such circumstances the aggregates are manufactured. Examples of artificial aggregates are broken brick, Blast furnace slag, M-sand etc. Broken Bricks are obtained from crushing over burnt bricks. Generally the crushing strength of the broken bricks obtained is less than 30 MPa and not suitable for RCC works. They can be used for PCC works and for the base of weathering course over the roof.

(b) **Classification Based on Size of Aggregates:** The sizes of aggregates vary from a few centimeters to microns. The size of aggregates should be chosen such that they fulfill the conditions that will be practical with regards to construction. Based on size the aggregates are classified as (i) Fine aggregates and (ii) Coarse aggregates.

(i) **Fine Aggregates:** Aggregates are classified as fine aggregates when the size of grains is less than 4.75 mm. Sand and silt are examples of fine aggregates. The size of sand ranges from 0.07 mm to 4.75 mm. The property of fine aggregate is to fill the voids between the coarse aggregates in concrete.

Sand: Sand can be obtained from natural sources like river banks or mechanically manufactured by crushing of rocks and gravels. According to IS383—1970, sand is classified into three type namely coarse, medium and fine sand.

(ii) **Coarse Aggregates:** Materials which are retained on 4.75 mm IS sieve (IS sieve no. 480) are known as coarse aggregates. Coarse aggregates are obtained from crushing of granites or bricks. The sizes of coarse aggregates used for construction start from 6 mm. Generally, for reinforced concrete structures, the size of coarse aggregate does not exceed 19 mm. For Plain cement concrete, for laying below foundation, the maximum size of aggregate is limited to 63 mm. For construction of concrete dam, where mass concreting is carried out the maximum size of aggregate can be up to 200 mm.

3.2.2 Properties of Aggregates

Aggregates have the following properties:

(a) **Bulk Density:** It is the mass of the aggregate per unit volume. Bulk density indicates the density of packing when filled in a standard way. The bulk density is dependent on four factors namely: (i) Particle size, (ii) Shape, (iii) Grading, and (iv) Moisture content.

A heavier bulk density indicates there are lesser voids to be filled by sand and cement in a concrete mixture. Generally bulk density should be more than 14.4 kN/m^3. Bulk density of aggregates has been specified in detail in IS2386 —1963.

(b) **Voids:** Voids are the empty spaces between the aggregate particles especially the coarse aggregates. Higher ratio of voids will result in weaker concrete and more consumption of finer particles. Void ratio is calculated as follows: $e = 1 - (y/Gs)$ where y is the bulk density and Gs is the specific gravity.

(c) **Particle Size of Aggregates:** According to IS456—2000 (pg 22), the nominal size of aggregate should not exceed one-fourth of the least dimension of the member. For RCC works, the maximum size of aggregates should be 20 mm.

(d) **Strength of Aggregates:** The strength of cement concrete is mainly dependent on the strength of coarse aggregates. Stronger the source of aggregates, higher the strength of concrete mix.

(e) **Specific Gravity of Aggregates:** Specific gravity of aggregate is defined as the ratio of the mass of solid in a given volume of sample to the mass of an equal volume of water at room temperature. Higher the specific gravity, heavier is the aggregate. Specific gravity is required to calculate the yield of concrete. The volume of concrete produced is depended mainly on the aggregates.

(f) **Shape and Texture:** Workability of concrete is dependent on shape and texture of aggregate.

- **Shape:** There are mainly four shapes of aggregates namely rounded, irregular, angular and flaky. Though rounded aggregates are easily workable, they reduce the strength of concrete. Angular ideals are ideal because they have a good workability and develop a good bonding which gives good strength to concrete.
- **Texture:** Surface texture is the smoothness of surface of aggregates and measured in the percentage. Texture are of following types:
  - Smooth: Marble, Slate
  - Glassy: Black flint
  - Granular: Sandstone
  - Crystalline: Granite, Basalt, Dolomite
  - Honeycomb: Trass.

  IS383 gives the relation between strength of concrete and texture of aggregates and the same is shown in Table 2.9.

Table 2.9: Table showing the relation between texture of aggregate and strength of concrete

Smoothness	Water/Cement	Compressive Strength	Tensile Strength
100%	0.54	34.8 MPa	4.3 MPa
50%	0.57	32.1 MPa	4.6 MPa
0%	0.60	29.5 MPa	4.8 MPa

(g) **Moisture Content:** The porosity of aggregates can result in moisture absorption due to which its actual weight may vary. Moisture content of aggregate has an effect on the water cement ratio and workability of concrete. There are mainly four conditions of aggregates in relation to moisture content namely dry, absorbed, saturated and free holding water. IS2386 describes the procedure to determine the moisture content in aggregates. The moisture content should be in the range of 0.5% to 1.0%.

(h) **Bulking of Fine Aggregates:** It is a phenomenon in which there is an increase in bulk volume of fine aggregates due to the presence of moisture. As an internal process, the bulking takes place when the sand particles are pushed by the film of water. Bulking of sand may vary from 15 to 30%. If the bulking of sand is found to be 40%, the sand is considered unsuitable for construction. IS2386 shows the graph showing the variation between bulking and moisture content.

3.3 Water

Water is a very important constituent of concrete mix. Water is required to initiate the chemical reaction to convert all the raw material and bind them to form concrete. It also plays an important role in strength. The water cement ratio affects both the strength of concrete and its workability. It is important that the quality of water is good and devoid of any harmful chemicals which will affect the strength and durability of concrete. The permissible limits of solids in water are specified in IS3025. Following are the requirements of water to be used for construction:

- The water should be potable.
- The water should be devoid of harmful chemicals like acid, alkalis, salt, sugar, organic materials and other substances deemed harmful for concrete and steel.
- The pH value should not be less than 6.
- Seawater is not recommended for production of concrete.
- The quality of water used for concreting and curing should be same.

3.4 Admixtures

Admixtures are artificial materials added to the concrete to enhance the properties of concrete to suit prevailing conditions. Admixtures are added to improve certain properties like workability, rate of hydration, durability, impermeability etc.

According to IS456, the admixtures may be administered to the concrete mix with the approval of engineer in charge for production of concrete. The approved admixtures have been detailed in IS9103.

Following are the functions of admixture:

- To improve workability of fresh concrete
- To retard the initial setting of concrete

- To accelerate the setting of concrete
- To increase the strength of concrete
- To increase the impermeability of concrete
- To increase the resistance to chemical attack
- To increase the bonding with reinforcement
- To increase the binding between old and new concrete
- To produce light weight or cellular concrete
- To produce coloured concrete.

3.4.1 Types of Admixtures

The various types of admixtures used are Accelerators, Retarders, Super plasticizers, Waterproofers, Bonding Agents, Pigments or coloring agents, Air entraining agents, Pozzolanic mixtures, and Grouters.

(a) **Accelerators:** As the name suggests, they accelerate the setting of concrete or in other words reduce the setting time of concrete. These are used when the structure has to be put to early use and hence the concrete has to set faster than normal. It is also used in cold climates where concrete takes time to set.

The most commonly used accelerators are calcium chloride. Addition of 2% of calcium chloride by weight of cement can reduce the setting time of concrete by 30%. Higher doses of calcium chloride can result in flash setting of concrete and increased shrinkage, hence they should be used up to a certain percentage only. IS7861 restricts the use of calcium chloride in cold weather conditions to a maximum of 1.5% by weight of cement. CaCl should not be used where there is high possibility of alkali aggregate reaction.

Alternatively, chemical compound of sodium like sodium sulphate, sodium nitrate etc. can also be used as accelerators. Excess use of sodium nitrate can cause cracking due to excess evolution of heat.

(b) **Retarders:** Doing the opposite of what accelerator does, it increases the setting time of concrete. This type admixture slows down the chemical process in concrete and helps keep the mixture plastic for a longer time. Retarders are used during extremely hot weather to prevent quick setting of concrete. They are also used where the concrete has to be transported to a longer distance. It is most commonly used when concreting for piles have to be carried out.

Generally gypsum is added in cement as it is a natural retarder. Common sugar can also be used as a retarding agent. The other examples of retarders are ammonium chloride, ferrous and ferric chloride, calcium porates, hydro-oxilated carbolic acid and lingo-sulphonic acid. Excess use of retarders will cause bleeding in concrete.

(c) **Super Plasticizers:** Workability is an important characteristic of fresh concrete. However increase in water content, even though it improves workability, reduces the strength of concrete. In case of high strength concrete, it is important to keep the water cement ratio in a controlled state. Due to increase in cement content, heat of hydration generated is high. Under these circumstances, super plasticizers are used to improve the workability of concrete. They can range from natural materials such as fly ash, bentonite clay, finely ground silica to artificial materials.

(d) **Water Proofing Agent:** Concrete by nature is not impervious and the voids present in concrete can aid in flow of water. To improve the impermeability, water proofers are added to the concrete during the production of concrete. This is done for water retaining structures such as underground sump and overhead tanks and members which have constant wet surface such as toilet slabs. The resistance of concrete to penetration of moisture can be improved by the addition of soda and potash soaps. Sodium silicate, zinc sulphate, aluminium and calcium chloride, petroleum wax etc. can also reduce permeability and surface staining and improve the water tightness of structure.

(e) **Bonding Agents:** This type of agent is generally used when, there are cold joints and new concrete has to be connected to the old one. Generally finely ground cast iron particles, polyvinyl acetate etc. are used as bonding agent. In design mix, the amount of agent is taken in the range of 5 – 20% by weight of cement.

(f) **Pigments or Colouring Agents:** These types of agents, as the name suggest are used to impart natural colour to the concrete. Pigments must be resistive to sunlight, must be chemically stable and not affect the major properties of concrete like strength and setting time. These agents are generally mixed with cement itself before the production of concrete. Generally metallic oxides are used as pigments. The mixing of red oxide will give red colour, chromium oxide or chromium hydroxide gives green, barium magnate or ultra marine give blue.

(g) **Air Entraining Agents:** This type of admixture helps in formation of air during the production of concrete. The air is formed in form of bubbles which fill the voids. The size of the bubbles range from 5μ to 8μ. These admixtures have following role to play in concrete:

- Improve workability and plasticity
- Improve durability
- Improve resistance to frost action
- Reduces bleeding and segregation
- Reduces permeability
- Reduces the weight of concrete.

(h) **Pozzolonic Mixtures:** Pozzolona is a silicious material found naturally. It can replace cement for a limited proportion (10–35% of cement by weight). The admixture based on pozzolona should confirm to IS9103. Following are the advantages of using this admixture:

- Lowers the cost of cement
- Lowers the heat of hydration and thermal shrinkage
- Improves workability
- Reduces bleeding and segregation
- Increases the resistance to attack from salt and sulphates
- Reduces the chances of alkali aggregate reaction.

However this admixture also results in decrease of gaining early strength which results in delay in removal of formwork.

(i) **Grouters:** Grouting agents are used to fill up the voids and cracks developed in hardened concrete. This is most common used for repairs. Generally methyl cellulose, starch etc. are the most commonly used grouters.

3.5 Properties of Concrete

As described above, concrete is a mixture of Cement, Aggregates, Water and sometimes admixtures. There are two stages of concrete namely fresh concrete and hardened concrete. Concrete, after production takes times to set. The concrete before setting is called fresh concrete and it is still in plastic stage and after setting is called hardened concrete.

3.5.1 Properties of Fresh Concrete

Fresh concrete is that stage of concrete when it is still in plastic stage. The fresh concrete should be homogenous and should not segregate or bleed. The main property of fresh concrete is workability.

(a) **Workability:** Workability is defined as the ease with which the concrete can be handled, transported, placed in position and compacted. According to IS1199, workability is that property of concrete which determines the internal work necessary for compaction. Following are the factors which affect workability:

  (i) **Mix Proportion:** Cement mortar in form of a paste is responsible for the cohesion of concrete. It acts like an adhesive and it creates the binding between aggregates. More the quantity of cement paste better will be the workability.

  (ii) **Water Content:** Water content has a major role to play in determining the workability of concrete. Increase in water content will increase the workability. However increase in water content also affects the strength of concrete. It can cause bleeding which will affect the binding of aggregates. Hence the concrete mix is designed for optimum water content which will give good workability and required strength.

(iii) **Properties of Aggregates:** Various properties of aggregates have their effects on the workability of concrete. They are shape and size, surface texture and grading.

- If the shape is rounded then, the concrete has better workability.
- If the surface texture is smooth then the workability is higher.
- Having equal sizes all through will create more voids and reduce the workability.

(iv) **Admixtures:** Admixtures like super plasticizers are used to improve workability. This is generally used in case of high strength concrete where the cement content is high and water cement ratio is low. These are also used in concreting of piles where the distance travelled is more and the concrete has to be self-compacted.

3.5.2 Properties of Hardened Concrete

The various properties of hardened concrete are as follows:

(a) Strength
(b) Durability
(c) Creep
(d) Shrinkage
(e) Thermal Properties
(f) Modulus of Elasticity

(a) **Strength:** The strength of concrete is generally referred to the compressive strength developed by the hardened concrete. It is also referred to as the grade of concrete. The permissible stresses are mentioned in IS456.

(b) **Durability:** Durability is the life of concrete for which it will be fit for service. The durability is dependent on the quality of ingredients, water-cement ratio, compaction of fresh concrete and curing of concrete.

(c) **Creep:** Creep is the strain developed in concrete due to stresses endured under prolonged loading. Like durability, creep is dependent on quality of concrete and stress developed in concrete. Additionally it is also dependent on the age of concrete. The permitted value of creep is 0.0003.

(d) **Shrinkage:** The reduction in size of concrete during the process of setting or hardening of concrete is known as shrinkage. Due to shrinkage, concrete experiences loss of strength. The permissible value of shrinkage is in the range of 0.02% to 0.01%.

(e) **Thermal Properties:** Concrete has very good fire resistive properties due to inherent presence of chemically bonded water. Additionally concrete contains pores or voids which create a vacuum which is a bad conductor of heat. Concrete in which broken bricks are used as aggregates have better fire resistive properties than the concrete in which broken stones are used as aggregates.

(f) **Modulus of Elasticity:** Elasticity is the property of a material by which it is returned to its original shape after the removal of applied stress. Mathematically modulus of elasticity is stress developed per unit strain. According to IS456, the value of modulus of elasticity of concrete is $5000\sqrt{f_{ck}}$ where f_{ck} is the strength or concrete in MPa.

4. STEEL

Concrete was found to be good in compression but weak in tension. To compensate the tensile properties it was decided to reinforce concrete with a material which had similar expansive properties. The concept of reinforced concrete using steel was invented by a French gardener, Joseph Monier, in 1849 and patented in 1867.

Steel is an alloy primarily made of Iron and has a carbon content of 0.2% to 2.1% by weight depending on the grade. Steel is primarily characterised by its ductility but the increase in carbon content increases the hardness and reduces ductility. The other advantages of steel are its weldability and resistance to corrosion.

The invention of Bessemer's process in the mid 19th century made the steel production cheap and facilitated mass production. Today steel is an integral part of the construction field. The ductile property of steel has helped in construction of High-Rise building. Till 1972, the tallest buildings in the world were made of steel framework. Today steel is widely used in industrial structures where the construction is fast and there is a scrap value when the building is no longer complies with serviceable criteria. It also enables re-usage with minimal wastage. The modern techniques to prevent rusting ensure that the shelf life, of steel the structures built with it, is increased considerably and thereby reducing the capital costs.

4.1 Physical Characteristics of Steel

The physical characteristics which make steel the ideal material to be used as reinforcement in Reinforced cement concrete are as below:

(a) **Coefficient of Thermal Expansion:** The coefficient of thermal expansion of steel is similar to that of concrete, which eliminates the large internal stresses due to thermal expansion or contraction.

(b) **Bonding:** The hardened concrete creates a good bonding with steel, which allows the structure to act as a homogeneous material.

(c) **Protection Against Corrosion:** Steel is corrosive and formation of rust decreases it's durability. The chemicals present in concrete form a film around the steel reinforcements which gives them a good protection against corrosion. Also the cover to reinforcement ensures that the steel doesn't come in contact with atmosphere.

The relative cross-sectional area of steel is in the range of 1% to 6% of the cross-sectional area of concrete.

4.2 Properties of Steel

The various properties of steel are as follows:

(a) **Elasticity:** Steel is a highly elastic material. The modulus of elasticity of steel is 2.0×10^5 N/mm^2.

(b) **Density:** Steel is a very dense material. Compared to water its density is closed to 8 times higher. The density of steel varies based on the alloying constituent but it generally varies from 77.5 to 80.5 kN/m^3.

(c) **Thermal Properties:** Steel has a very low thermal conductivity. Value of thermal conductivity of mild steel is in the range of 26 to 375.5 BTU.

(d) **Strength:** The various properties of strength for steel are:

(i) **Tensile Strength:** Steel is very good in tension. The tensile strength of steel is taken as a percentage of yield strength. Tensile strength of mild steel is 247 MPa.

(ii) **Bond Strength:** The amount of adhesion between bonded surfaces is measured in terms of the stress required to separate a layer of material from the base to which it is bonded. Bond strength is a measure of the transfer of load between concrete and steel. Tor steel has a better bond strength with concrete than the mild steel.

(iii) **Toughness:** Toughness is the ability of a material to absorb energy and plastically deform without fracturing. Toughness in steel depends on the carbon content. Higher percentage of carbon content in steel makes it brittle.

(e) **Ductility:** Ductility is the ability of the material to be drawing into thin wires. Steel is a highly ductile material. Due to this steel can be drawn into thin wires which are used as binding wires and in wire mesh.

4.3 Types of Steel in Construction

The different types of steel used in construction are as below:

(a) **Structural Steel:** The structural steel is the type of steel which is used directly as the main material in the structural form. All the industrial buildings are made of steel. The advantage of this is the speed in which the construction can be carried out and the lower cost of construction. The use of cold rolled sections enables the reduction in section thickness and subsequently the reduction in cost. The various shapes of structural steel are I sections, Box sections, Angles, Channel, Circular pipes, Plates, Z sections etc.

(b) **Reinforcement Steel:** The steel used for reinforcement in concrete are of two types: Mild Steel rods and HYSD rods. Mild Steel contains carbon upto 0.23 to 0.25%. Higher value is permitted for bars of diameter of 20 mm and more. The grade of steel is Fe250 which indicates that it is yield strength is 250 N/mm^2. Nowadays TOR steel is replacing it and use of mild steel is restricted to window bars for grills and other non-structural uses like tamping rods.

HYSD (High Yield Strength Deformed) rods are used for its higher strength both in tension and in bonding. The grade of steel is Fe415, Fe500 and Fe550 where the number denotes the yield strength. They have higher carbon content and sometimes the larger diameter rods are brittle in nature. To improve the ductility the carbon content is reduced without affecting the yield strength.

(c) **Steel Wires**: Steel is a highly ductile material and hence it can be drawn into thin wires with ease. Steel wires are widely used for various works in construction. They are used to bind reinforcement placed for RCC works. They are also used in form of wired mesh. Steel wire mesh is widely used in Ferro-cement works, in steel windows for protection. High tensile steel wire or strands are used in pre-stressed concrete.

5. PRESTRESSED CONCRETE

Cement concrete has good compressive strength but weak in tension. To enhance the tensile properties of a section, steel is reinforced in tension zone. So effectively concrete takes care of compression and steel takes care of tension and together it is called reinforced cement concrete (RCC). However the life of a RCC section is limited in the sense that once the steel starts getting corroded, the strength come down drastically as corrosion is not a linear process. To overcome this drawback, a process was developed to introduce the stresses in concrete prior to its loading to ensure the tensile stresses are taken care. By definition pre-stressed concrete is "structural concrete in which internal compressive stresses have been introduced to reduce potential tensile stresses in the concrete resulting from loads". This covers for the deterioration of concrete due to corrosion of reinforcement.

5.1 History of Prestressing

The concept of prestressing is as old as the civilization. There have been records of Egyptians using this concept initially in their boats where the system of post and rope was used to prevent hogging of hull.

In modern world the concept of prestressing was first patented by P.H. Jackson in 1888 in the United States. His idea of using high strength steel which exhibited low relaxation characteristics was perfect but not available at that time of patenting the design. It was not until a few years later, upon the French structural engineer Eugene Freyssinet's proposals of using high strength steel, prestressing became practically viable. However it would take Gustave Magnel to propagate the concept of design of prestressing. His book, *Le Béton Précontraint*, was the first book written on this subject.

In 1950's TY Lin popularized the concept of prestressing by standardizing the practical use of prestressed concrete. He emphasized the need for structural aesthetics in engineering regardless of the economy of the project.

5.2 Advantages and Disadvantages

Prestressing has the following advantages:

- It has a longer life than RCC structure.
- Sections can be designed more economical than RCC structure for the same span and load. However due the equipment cost, it economical only for larger spans and larger constructions.
- Once the stresses are transferred, the steel does not have a role to play.
- It has lesser maintenance cost dues to repairs.

It has the following disadvantages:

- The cost of equipment and set is high thus making it uneconomical for smaller constructions.
- It is difficult to have variance in profile.

5.3 Types of Prestressing

There are mainly two types of process of prestressing, namely Pre-tensioning and Post-tensioning.

5.3.1 Pre-tensioning

As the name suggests, in this method, the tendons are pre-tensioned before the concreting is carried out. In this method, the tendons are placed in position and imparted tensile stress. Then the concrete is poured and allowed to set. After the concrete is set, the tendons are released from the end jacks and the stress is transferred to the concrete.

5.3.2 Post-tensioning

In this process, the tendons used are tensioned after the concrete is set. In this method, the tendons are placed in position in a duct and then the concreting is done. After the concrete is set, the tendons are imparted tensile stress. The duct is grouted with concrete and when the grout is sit, the tendons are released from the jack and the stress is transferred to the concrete.

6. FINISHING MATERIALS

Finishing of a building is a very important aspect that makes the structure pleasing to both onlookers and user. It is like clothing a structure. The different kinds of finishing materials are as follows:

6.1 Flooring and Cladding

To improve the look and feel of floor and wall, various types of materials are used. The various types of material used for flooring and cladding are as follows:

(a) **Natural Stones:** Natural stones such as granite, marble, slate are used as flooring and cladding materials. These are procured naturally, processed and finished based on requirement. Earlier they were available abundantly and hence the cost of laying natural flooring was low. Furthermore there was no constraint on the completion time of the project as laying of natural stones is labour intensive. However in today's scenario the cost of procuring the natural raw material is high and the time constraint in most cases make it an expensive affair. Examples of natural stones are Granite, Marble, Shahabad, Kota, Cuduppah.

(b) **Oxide Flooring:** In earlier days, the flooring especially in houses was made of cement which was not aesthetically pleasing. Furthermore the cement flooring increased the temperature inside the room due to its inherent properties. To improve aesthetics and reduce the temperature, oxides were mixed to prepare the flooring. The most common oxide flooring was the red-oxide flooring which gave a good finish and a good cooling effect. However it is labour intensive and time consuming and in today's day where both labour and time is rare commodity, it is not preferred.

(c) **Clay Tiles:** It is a type of material made of burnt clay. It is also known as terracotta tile. Before the popular use of concrete as a structural material, clay tiles were used for both flooring as well as roofing. The tile used for flooring is flat whereas the ones used for roof comes in different shapes and sizes. The ones used for terrace is also known as weathering tiles.

(d) **Ceramic Tiles**: Ceramic tiles are the most popular flooring and cladding material in today's construction. All the toilets and kitchen walls today are cladded with ceramic tiles. The advantages of ceramic tiles are:

- It is inexpensive
- It is durable
- It can be laid very fast
- It is easy to maintain
- It does not absorb water

Fig. 2.3(a): Granite flooring

Fig. 2.3(b): Oxide flooring

Fig. 2.3(c): Clay tiles

Fig. 2.3(d): Ceramic tiles

Fig. 2.3(e): Polymer flooring

Fig. 2.3: Figure showing various flooring material

Prior to 20th century, ceramic tiles were expensive material and used to adorn public buildings and homes of the wealthy. The history of decorative tiles dates back to 4000 BC and is found in the Egyptian civilization. In the early middle ages, the Chinese popularized the concept of porcelain art and widely used it for decoration. The concept of ceramic tile became widespread and each civilization customized the art according to their culture.

The industrialization changed the way the tiles were manufactured. Mass production of tiles became a reality and that made ceramic tiles affordable and widely used. The latter half of 20th century saw major advancements and innovations in tile industry. Digital technology and precision equipments have revolutionized the design and manufacturing of tiles.

(e) **Artificially Processed Polymer Flooring:** Polymers like PVC have been used to make flooring materials. One of the most common examples is vinyl flooring. They come in form of rolled sheet or tiles. The advantage of this type of flooring is the low cost and the speed of fixing. However it has a very low shelf life.

(f) **Glass Tiles:** Glass tiles are relatively new material for flooring and cladding. The original use of glass tile was carried out in swimming pools. Glass mosaics were used as cladding material for the swimming pool. They were also used as cladding materials in external walls and in toilet walls. However it is a very labour intensive work and it does get expensive.

6.2 Painting

Paint is a coloured substance which is spread over a surface to give a decorative and protective coating. By definition, paint is any liquid, liquefiable material or mastic composition which after application forms a thin layer on the surface. It is used to protect, color or provide texture to the surface it is being applied on. It can be considered as giving a final makeup to the object it is applied on.

The various types of paints are:

(a) **Cement Paint:** Cement paint is a material which consists of OPC and other chemical ingredients which add colour to the mixture. The material is water based and is in form of a powder. This type of paint is applied to the walls and very good as exterior paint. It is described in IS5410.

(b) **Oil Based Distemper:** It is a type of water based paint which was manufactured by mixing oil. Originally the distemper was widely used in paintings and the binder included glues of vegetable or animal origin. This type of paint is used for interior walls and the biggest advantage of this type of wall is the cleaning of the surface. Another advantage is the surface repels water due to the oil present in the mixture.

(c) **Acrylic Emulsion:** Acrylic is an artificial pigment which is used in manufacture of water based paint. Though acrylics are water soluble, they become water resistant after drying. The difference between oil based distemper and acrylic emulsion is the binder used in paint and time taken to dry.

(d) **Varnish:** Varnish is a type of paint used for wood. Varnish is a hard protective finish and is a combination of drying oil, resin and a thinner or solvent. Varnish has no color and is usually glossy but can also be designed to produce semi gloss or satin finish.

6.3 Joineries

The different kinds of joineries are doors, windows and ventilators. The various materials used in manufacture of joineries are as follows:

(a) **Timber:** Timber was the earliest form of material used as joineries. It is naturally available and durable due its seasoning with local climate. The most common wood to be used as joinery was Teak wood as it was durable and long-lasting and it gave a rich look when finished. However due to its cost, other timber materials, which were more locally available, were used for joineries. These woods were padak, kongu, sal, neem etc. The main doors however continue to use teak as the material.

(b) **Steel:** Steel joineries are mainly used in industrial structure because of its durability and cost. Generally factories need not have aesthetics and the clients prefer low cost materials. Hence the joineries especially windows are made of steel sections which are strong and have a long life. The drawback is that there needs to be a periodic maintenance otherwise they can get rusted.

The manufacture of stainless steel and increasing cost of timber has made the engineers consider manufacturing steel doors. Technology has facilitated the improvement of aesthetics too where a steel door can be manufactured to give a wooden finish.

(c) **Aluminium:** Increasing cost of timber, its durability and time of manufacturing have made engineers look for alternative material. Aluminium being abundantly available began to replace wood for windows. The aluminium windows were manufactured using aluminium as frames and glass panels. The advantages of the aluminium windows were:

- It is abundantly available
- It is cheap
- It does not rust
- It does not deform due to climatic conditions especially with respect to absorption of moisture.

Initially aluminium was used in its natural colour which looked cheap. Subsequently the process of powder coating and anodisation made it possible to give variety of colours to the section and make it look good and aesthetic.

(d) **Poly Vinyl Chloride or PVC:** The first use of PVC was made in doors for toilets. Wood absorbed moisture and hence less durable. To eliminate this drawback, PVC was used for toilet doors. The advantage of these doors is that it is light, cheap and does not get affected by moisture. However the drawback was that it was not suitable for external use where there was exposure to direct sunlight.

The development of UPVC (unplasticized PVC) enabled it to be used for windows too. Originally the cost of UPVC windows was high and used only where there was no restriction on the cost of construction and durability was required. However with the ever increasing cost of timber and the labour for manufacturing wooden windows, UPVC windows have become popular and widely used.

Initially PVC doors were made of extruded hollow strips stiffened in the edges. Improvement of technology enabled the manufacture of paneled doors. In the initial days PVC was produced with colour choices. The most common UPVC windows were manufactured in white. Today the colour and texture of the doors and windows are such that they can be matched to make them look like wood.

(e) **Concrete:** Concrete joineries were experimented for the manufacture of low cost joineries. The frames were prefabricated and the internal panels were either glass or steel. Due to its difficulty in handling this type of joinery is not popular. However prefabricated concrete jallys are used for ventilators and in low cost constructions.

Reference and Further Reads

Duggal, S.K., "Building Materials", 4th Edition, 2016, New Age International (P) Limited.

Varghese, P.C., "Building Materials", 8th Edition, 2012, Prentice Hall International.

Ward-Harey, Ken, "Fundamental Building Materials", 4th Edition, 2009, Universal-Publishers.

Zhang, Haimei, "Building Materials in Civil Engineering", 2011, Woodhead Publishing.

3 STRUCTURE

Structure is any construction or framework of identifiable elements or components which give form and stability and is suitable for the use it is intended and constructed. The form of the structure takes care of the utility and aesthetics. The stability of the structure is its resistance to withstand various loads and stresses it will experience during its lifetime. It is analogous to the skeleton of the human body. The structures have definite boundaries and are confined within a limited space. The structures can be one dimensional, two-dimensional or three-dimensional.

Life of structure is defined as the period of time for which it will serve the purpose for which it is originally built. Life of the structure is determined based on the requirements of use. It is dependent on the material used for construction. Structures made of artificial material have limited lifespan. An RCC structure has a lifespan of 50 – 100 years. Structures made of natural material have stood for centuries.

1. CLASSIFICATION OF STRUCTURE

There are various forms of classification of structures based on different criteria. The broad classifications of structures are as below:

1.1 Classification Based on Occupation

Based on occupation, the structure can be classified as:

(a) **Habitable:** Habitable structures are those types of structures which are occupied by humans for various purposes. Buildings are a type of habitable structure. Generally the word habitable is used more in a residential context. A space in a building used for living, sleeping, eating or cooking are termed habitable spaces. So effectively, bathrooms or toilets are non-habitable spaces.

(b) **Non-habitable:** Non-habitable structures are those types of structures where the above activities in not possible. Structures like towers, silos, dams and bridges are non-habitable structures.

1.2 Classification Based on Height

Based on height, the structure can be classified as:

(a) **Short Structures:** From the design consideration, a structure is termed short structure if the vertical loads are predominant and the effects of horizontal forces are negligible. Generally any structure less than 10 m in height is considered as short structure.

(b) **Tall Structures:** Conversely in a tall structure, the horizontal forces like wind or earthquake become more prominent and the design criteria are based on these loads. Tall structure is dealt in detail in chapter 9.

1.3 Classification Based on Location

Based on location, the structure can be classified as:

(a) **On Shore Structure:** Structures built on land are called as on shore structures. Generally buildings, towers, etc. are on shore structures.

(b) **Offshore Structure:** Offshore structures are those structures which are built away from land. Structures like ports, bridges, oil wells in water are few examples of offshore structures.

2. COMPONENTS OF A STRUCTURE

Any structure has two components—sub-structure and superstructure. Sub-structure is that portion of the structure which is generally constructed below the natural ground level. Conversely, superstructure is that portion of the structure which is constructed above the ground.

2.1 Sub-Structure

A sub-structure is that part of the building which is constructed below the natural ground level. The most common sub-structure is the foundation.

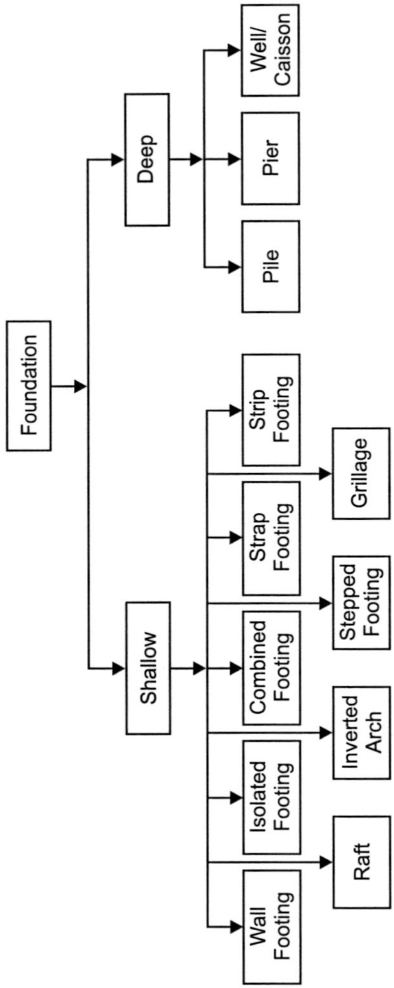

Fig. 3.1: Figure showing classification of footings

Foundation is the most important part of any structure. It is that part of the structure which is in direct contact with the soil. The surface on which the foundation rests is known as the foundation bed or sub-soil. It is important to ensure that the foundation is designed and executed properly as it is very difficult and expensive to alter the foundation once constructed.

The various functions of the foundation are:

- Transfer load to soil safely.
- Support the superstructure.
- Ensure stability of structure against both vertical as well as horizontal load.
- It has to be designed in such a way so that there is no differential settlement.

Differential settlement is a condition in which one column or one side of column or row settles unevenly from the other columns. The differential settlement takes place due to:

- Poor sub-soil, namely when soil has excessive shrinking or expansive properties.
- Excessive movement of groundwater and saturation of soil.
- Excessive vibration transmitted to sub-soil.
- Slipping of uneven sub-strata.

It is a common notion that the foundation has to be placed very deep for better load bearing capacity. However that is not true. In the case foundation on rock, it is not possible to carry out excavation. In olden days it was not possible to improve the strength of the soil or natural foundation and hence the excavation was carried out to the point where the bearing capacity was found to be good. However with the advancement in the field of soil mechanics and introduction of new techniques in design and construction it is possible to place the foundation at a lesser depth. The minimum depth at which the foundation can be placed is 900 mm. The foundation can be divided into two categories namely shallow foundation and deep foundation.

2.1.1 Shallow Foundation

Shallow foundation is that type of foundation where the depth of the foundation is generally less than the width of the foundation. In general the procedure to construct the foundation is preceded by the excavation of soil, creating an even surface, stabilizing of soil if required, placing of reinforcement as per design and then pouring of concrete in desired shape. The various types of shallow foundations are Wall footing, Isolated footing, Combined footing, Strap footing, Strip footing, Raft, Inverted arch, Stepped footing, and Grillage.

(a) **Wall Footing:** Wall footing is a type of continuous footing designed for load bearing structures. In this case the footing runs all through below the wall. The shape of the foundation is in form of steps on both sides of the wall. The width of the footing is generally more than 2 times the width of the wall it is supporting.

(b) **Isolated Footing:** Isolated footing is a RCC footing, provided for framed structure. It is the simplest form of footing for a framed structure. Such a footing is provided for each column and it is monolithically cast with column. Isolated footing is preferred for its low cost and it the preferred choice of footing, where there is no restriction of space. To reduce the cost, the footing is sloped towards the edges.

(c) **Combined Footing:** Combined footing are provided where there is a restriction on space for providing isolated footing or the columns are spaced closely and the isolated footings overlap. This type of footing can also be provided where there is uneven loading pattern on two adjacent columns and the sizes of the isolated footings vary.

(d) **Strap Footing:** Strap footing is a type of footing where the two isolated footings are connected with the help of strap beam. Strap beam is provided where either one or both the footings are at the edge of property line and there is an eccentricity in the isolated footing. By providing the strap beam, the additional moment coming on the footing due to eccentricity can be avoided and the size of the footing can be limited.

(e) **Strip Footing:** Strip footing is a type of combined footing. It is adopted for more than two columns placed so close that providing an isolated footing is not possible. The shape of the footing would be like an inverted T where the foundation slab is in contact with the sub-soil and the columns are connected by a strip beam. The advantage of the strip footing is that in case of uneven loading, on a set of columns, the footing distributes the load evenly and transfers to the sub-soil uniformly.

(f) **Raft:** Raft foundation is provided whenever the area required by the column load is more than half the total available land area available for foundation. Raft foundation is provided for the entire area connecting all the columns like a mat and it is also called mat foundation. It is a good foundation to where:

- The soil is poor and having low SBC.
- There is a chance of excessive differential settlement.
- The load on superstructure is high.

(g) **Inverted Arch Footing:** In earlier days where the load was heavy and the normal footing was insufficient to take the load, this type of footing was adopted. Generally this type of flooring is adopted for piers. The arch shape can reduce the intensity of superimposed load to a great extent. This happens because there is a negative moment developed due to the horizontal stresses developed in the arch. The negative moment counters the positive moment and thus reduces the bending moment developed due to super imposed load. However the subsoil needs to have good bearing strength to construct this type of footing.

(h) **Stepped Footing:** This type of footing is adopted when the structure is built on a gradient or slope. In slopes it would be uneconomical to carry out the entire excavation. Furthermore it may not be practically possible to excavate too. This is generally the case in hills. Is such a case, foundation is constructed in form of steps where each layer overlaps the other and transfers the load. The minimum depth of the foundation should be 800 mm and the overlap should be twice the thickness of steps. To control or prevent scouring of soil at toe level, the minimum clear distance between the existing ground and the step should be 600 mm where the sub-soil is rock and 1000 mm in all other cases.

(i) **Grillage:** This type of foundation is provided for steel columns and generally used in industrial structures. In industrial structures, the steel columns supporting gantry girders are subjected to heavy rolling and impact load. The foundation experiences vibration and concrete footing may be insufficient to withstand such loading. Under such circumstances, grillage foundation is provided which a steel foundation consisting of I sections as top and bottom tiers encased in concrete for protection.

2.1.2 Deep Foundation

Deep foundation is that type of foundation where the depth of the foundation is far greater than the width of the foundation. The various types of deep foundations are Pile foundation, Pier foundation, and Well or caisson foundation.

2.1.2.1 Pile Foundation

Piles are needle like structures, driven into the ground and resting on a hard stratum and transfer the load of superstructure to sub-soil safely. The transfer of loads can be either through friction or through end bearing. Piling is not a new concept and historically speaking there is evidence that the piles have been used even 4000 years ago on the shores of Lake Geneva opposite Morges in Ireland. In India too, many structures have been supported on piles and for the structures built during the mughal period. The famous Taj Mahal built in 1632 is founded on wooden piles. Pile foundation can be adopted under the following circumstances:

- When the available sub-soil is poor like black cotton soil or the soil is saturated.
- When the load on superstructure is high and design of shallow foundation is insufficient.

- When the structure is constructed offshore either in sea or river.
- When the excavation of soil is difficult and expensive.

The classification of piles based on various considerations is shown in figure 3.2. Primarily the piles are divided into two categories namely (a) load bearing piles and (b) non-load bearing piles.

(a) **Load Bearing Piles:** As the name suggests the load bearing piles are those which carry load and transfer load from the superstructure to the subsoil. Generally the pile is driven to a point where there is a hard stratum but sometimes when the containing soil has enough friction then the pile is designed accordingly. Load bearing piles are further classified according to mechanism of load transfer and according to material.

   (i) According to mechanism of load transfer the piles can be classified as the following:

   - **End Bearing Pile:** In this type of pile, the pile is driven to such a depth so that it rests on hard stratum. This is an ideal condition of foundation because of it stability and high SBC at a greater depth.
   - **Friction Pile:** In certain cases, hard stratum might be very deep and driving piles to that depth might become very expensive. This type of pile is adopted if it is found that the surrounding soil has enough cohesion to develop the friction required to withstand the load on piles. The pile is then designed based on the criteria of friction encountered on the surface.
   - **Batter Pile:** Sometimes where the foundation experiences very heavy horizontal load, then the piles are driven in an inclined position to counter both the horizontal as well as vertical load. Such type of piles are known as batter.

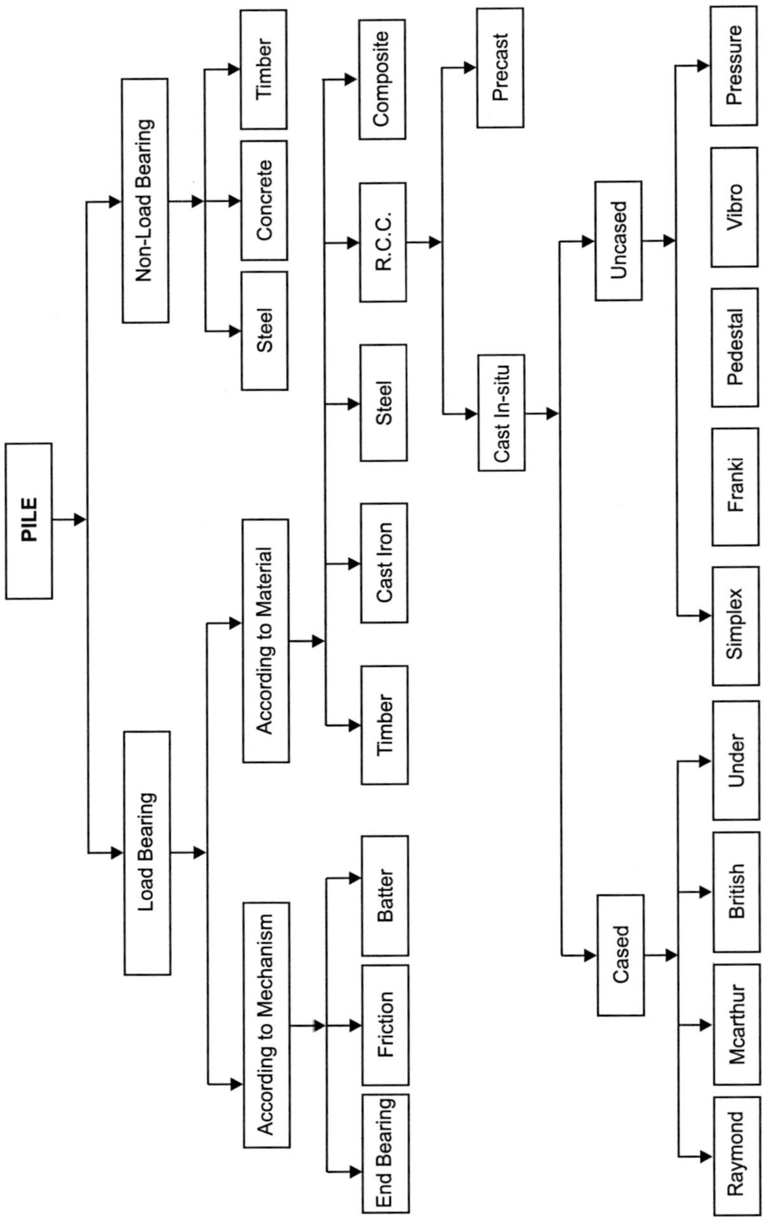

Fig. 3.2: Figure showing classification of piles

(ii) According to the type of material the piles are classified as following:

- **Sand:** Sand piles or more commonly known as the Sand Compaction Pile (SCP) method is used frequently in construction to form compacted sand piles by vibration, dynamic impact or static excitation in soft ground. Originally developed in Japan to improve stability or compressibility and to prevent liquefaction failure in loose sand, the SCP method is now often applied to soft clay ground to ensure stability and reduce ground settlement.
- **Cast Iron:** Ductile Iron Piles are an innovative deep foundation solution that can address a wide-range of geotechnical and construction challenges. With installations to treat a variety of soil conditions including fill, organics, soft clay and loose sand, Ductile Iron Piles can develop high capacities in either end-bearing or by developing a grouted bond zone to transfer loads in friction. The system is installed with medium-sized excavators for easy mobilization to a project site. The modular 5-meter long pile sections are easy to transport and move around tight sites. Installation using high frequency percussion hammers results in low vibrations, allowing the system to be used in close proximity to structures. The advantage in using this type of pile is the speed of construction.
- **Steel:** The use of steel has been increasing ever since the steel was manufactured. The main reason for use of steel is the high resistive strength, ease of handling and the variety of shapes. Steel piles have the ability to withstand not only vertical load but horizontal load too. The wide use of sheet pile in excavations is a testimony to this.
- **Timber:** The earliest piles were made of timber as it is a naturally available material. The timber piles are made from timber obtained from trees like sal, teak, deodar, babul, khair etc. It has been found that piles

made from Khair wood can stand action of sea water better and are thus commonly used for marine works. Timber piles prove economical hit supporting light structures to be located in compressive soils constantly saturated with water.

- **Composite:** As the word describes, composite piles are piles made of two or more material. Generally it is a combination of one pile driven over another. In such a combination, advantage is taken of the good qualities of both the materials. These prove economical as they permit the utilization of the great corrosion resistance property of one material with the cheapness or strength of the other. The common combinations are concrete over timber or concrete over steel.
- **Concrete:** Concrete piles are the most commonly used material for the construction of piles because of low cost, low maintenance and ease of construction. Concrete piles can be either cast-in-situ or precast. The cast-in-situ piles are further classified as cased concrete cast-in-situ piles and uncase cast-in-situ concrete piles.

(b) **Non-Load Bearing Piles:** Non-load bearing piles do not carry any structural load. They are generally not a part of the building. They are extensively used to prevent soil sliding during construction of foundations, dams retaining walls etc. They have the following functions:

- To isolate foundation from adjacent soil.
- To prevent underground movement of water.
- To prevent transfer of machine vibration to adjacent structure.
- To prevent soil sliding during construction of retaining wall.
- To protect river banks.

These types of piles can be made of timber, steel or concrete. Sheet piles are an example of non-load bearing piles.

2.1.2.2 Pier Foundation

A Pier foundation consists of a cylindrical column of large diameter to support and transfer large super-imposed loads to the firm strata below. The difference between pile foundation and pier foundation lies in the method of construction. Though pile foundations transfer the load through friction and/or bearing, pier foundations transfer the load only through bearing. Generally, pier foundation is shallower in depth than the pile foundation. Pier foundation is preferred in a location where the top strata consist of decomposed rock overlying strata of sound rock. In such a condition, it becomes difficult to drive the bearing piles through decomposed rock. In the case of stiff clays, which offer large resistance to the driving of a bearing pile, pier foundation can be conveniently constructed. Pier foundations are used for offshore structures and structures on water like bridges and ports.

2.1.2.3 Well Foundation or Caisson

A caisson foundation also called as pier foundation is a watertight retaining structure used as a bridge pier, in the construction of offshore structures. It is a prefabricated hollow box or cylinder sunk into the ground to some desired depth and then filled with concrete thus forming a foundation.

Caisson foundation is used in the construction of bridge piers and other structures that require foundation beneath rivers and other bodies of water. Caisson can be floated to the job site and sunk into place. Caisson foundation is similar in form to pile foundations, but is installed using a different method. It is used when soil of adequate bearing strength is found below surface layers of weak materials such as fill or peat. It is a form of deep foundation which is constructed above ground level, then sunk to the required level by excavating or dredging material from within the caisson.

Caissons (also sometimes called "piers") are created by auguring a deep hole into the ground, and then filling it with concrete. Steel reinforcement is sometimes utilized for a portion of the length of the caisson. Caissons are drilled either to bedrock (called "rock caissons") or deep into the underlying soil strata if a geotechnical engineer finds the soil suitable to carry the building load. When caissons rest on soil, they are generally "belled" at the bottom to spread the load over a wider area. Special drilling bits are used to remove the soil for these "belled caissons".

2.2 Superstructure

A superstructure is that portion of the structure which is constructed above the ground level. The loads coming on the superstructure is transferred to the foundation which in turns transfers to the subsoil safely. A superstructure is either a load bearing structure or framed structure depending on the type of building. In earlier days most of the structures built were load-bearing structures. However a lot of temples built around south India did have framed structures built out of stone in certain portions of the temple. In today's scenario, due to the changes in geotechnical considerations, complexity of structures and time factor, load bearing structures are almost obsolete.

2.2.1 Load Bearing Structures

In load bearing structures, the roof or floor load is transferred to the walls it is being supported on and the walls transfer the load directly to the foundation. Till around 40 years ago, most of the structures up to three stories were built as load bearing structure considering the cost factor. The disadvantage of such structure is that, the height of the structure was limited and taller the structures, the bigger were the size of the walls. The openings for doors and windows had to be planned in advance. Once constructed, they could not be altered or changed. Even the internal walls of a building were designed and constructed as load bearing to reduce the span of the slab and to limit the size of the foundation.

2.2.2 Framed Structures

With the advent of wide usage of Reinforced Cement Concrete and the changes in geology, the concept of building structures changed to framed structures. Due the increase in the activity of natural forces like earthquakes, the structures needed to be more stable and be able to withstand the natural forces.

Framed structures are a system of vertical and horizontal members connected in different forms and act monolithically to resist all the forces coming on the structure. The horizontal members are beams and slabs and the vertical members are the columns. The advantage of this system is that, the thickness of the walls could be reduced because the structure is designed in such a way that all the loads are resisted by the frame and walls are just fillers.

In a framed structure, the dead load and live load is transferred to the slab. The slab in turn transfers the load to the beams on which they are supported. The beams then transfer the load to the columns and finally the columns transfers the load to the foundation where the load is dissipated to the ground safely. In addition to the gravity load, in case of multi storied structure, the structure has to withstand horizontal loads too namely, wind and earthquake. Because of the monolithic nature of the structure, the framed structures are better equipped to withstand horizontal loads.

The various components of a framed structure are as follows:

(a) **Columns**: Columns are vertical members which predominantly withstand vertical load transferred from the structure they are supporting. All the loads coming on slabs and beams are transferred to column which them transfer the load to foundation. In high rise buildings, columns also have to withstand horizontal load experienced by the structure.

Fig. 3.3: Figure showing various structural elements

Generally column is designed as compression member. Columns can be classified as long column and short column based on the ratio its height to least lateral dimension. If the ratio of its height (unsupported length) to least lateral dimension is less than 12, then it's a short column. If the value is greater than 12, then it's a long column.

(b) Beams: Beams are horizontal members which are designed to withstand loads transferred from slab. They also take care of wall load which is directly resting on the beam. A beam is designed as a bending member. Beams can be classified depending on the end condition of support and depending on the shape of the cross-section of design.

(i) Depending on the end condition of the supports the beams are designed as:

- **Simply Supported:** Simply supported beams are those types of beams in which the end supports are pinned or rollers. The typical example of pinned joint is a beam placed on column with a pad placed in between them. This is seen generally in over bridges constructed.
- **Fixed:** Fixed ends are those types of supports where the joints are rigidly connected. Typical example of fixed end is a framed structure where the junction of beams and column are rigid and monolithic. Both simply supported and fixed beams are designed for a single span beam.
- **Continuous:** Continuous beams are those types of beams which have multiple spans. The end supports of a continuous beam can either simply supported or fixed.
- **Cantilever:** Cantilever beams are those types of beams which can are supported only on one end and free on the other end. The supported end is a fixed end. One of the best examples of cantilevered structure is human arm.

(ii) Depending on the shape of cross-section, beams are designed as the following:

- Rectangular beam.
- T – Beam
- L – Beam

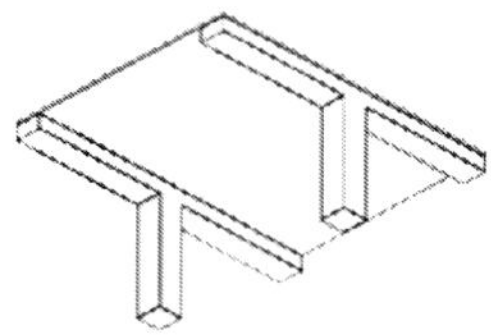

Fig. 3.4(a): One-way slab

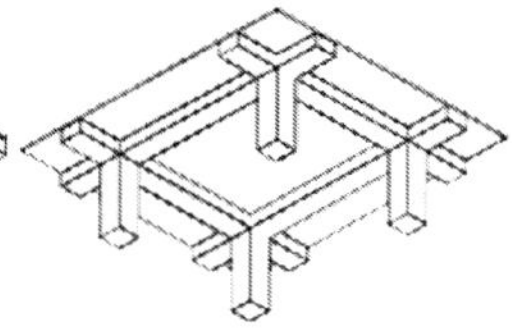

Fig. 3.4(b): Two-way slab

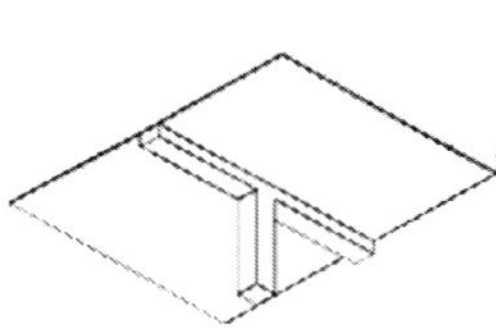

Fig. 3.4(c): Cantilevered slab

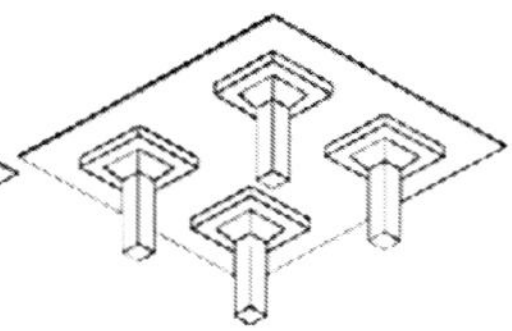

Fig. 3.4(d): Flat slab

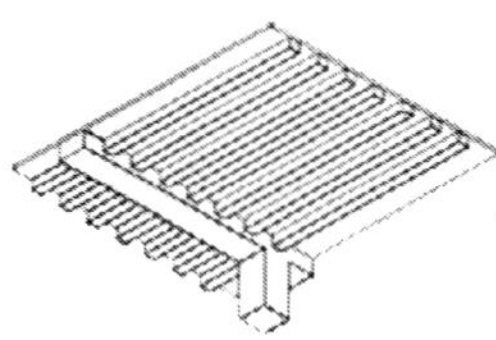

Fig. 3.4(e): Ribbed slab

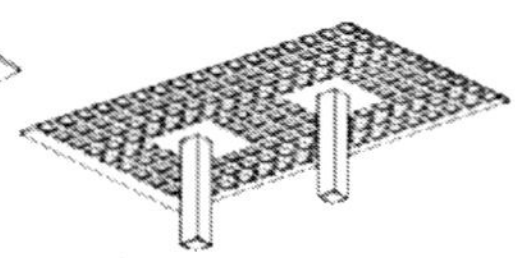

Fig. 3.4(f): Waffle slab

Fig. 3.4: Figure showing different types of slabs

(c) **Slabs:** Slabs are horizontal members or plate like structures that withstand dead loads and live loads. The load on a structure is transmitted to the slab directly. Slab is designed as bending member. Depending on the end conditions slabs are categorized as follows:

- **One-way Slab:** One-way slab is that type of slab where the load is distributed only to two ends of the slab. By definition one slab is that type of slab where either the slab is supported only on two sides or the ratio of longer side to shorter side is greater than 2. The stresses are distributed only along one direction.
- **Two-way Slab:** Two-way slab is that type of slab where the ratio of longer side to shorter side is less than 2 and all four sides are supported. The load is distributed on all sides and stresses are distributed on both directions of the plane.

- **Cantilevered Slab:** It is similar to a cantilever beam.
- **Flat Slab:** Flat slab is a type of slab supported directly by concrete columns without the use of beams. Flat slab is defined as one sided or two-sided support system with shear load of the slab being concentrated on the supporting columns and a square slab called 'drop panels'.
- **Ribbed or Waffle Slab:** Ribbed slab or waffle as it is called is a type of slab which is a combination of slabs and rib beams placed at very close interval. Both the slab on top and rib beams at the bottom is thin sections. Ribbed slab is a type of slab where the rib beams are placed in one direction and waffle is a type of slab where the rib beams are placed in both directions. The advantage of this type of slab is its higher load carrying capacity with slender sections. However due to the disadvantage in the cost of formwork and time taken to erect the formwork, it is used in large structures where the formwork cycle is repeated. It is ideally used for car parking structures. In earlier days this was a common type of roof slab with wooden rafter being used as ribs and the clay tiles used as the roof covering. Madras terrace is an example of ribbed slab.

(d) **Arches:** Arch is an element which has a curved profile. Arch is a pure compression member. The forces experienced by arch are resolved into compressive stresses, thus eliminating tension. This was called arch action and it was the most popular form of member for large spans as the materials used for construction in the earlier days had low tensile strength.

Based on the support conditions, arches are classified as fixed arch, two hinged arch and three hinged arch. The two hinged arch is most often in bridges of long span.

Based in the form or shape, arches can be classified as circular, pointed and parabolic.

Fig. 3.5(a): Parabolic and Circular arch

Fig. 3.5(b): Pointed arch

Fig. 3.5(c): Arch Bridge

Fig. 3.5: Figure showing types of Arches

(e) **Concrete Walls:** Concrete walls are vertical members which are constructed mainly to withstand horizontal load on the surface of the wall. They are constructed above the ground as a structural system for tall buildings which are termed as shear walls or below ground for retaining structures like underground sump, or retaining walls. The shear walls are designed to withstand both gravity as well as horizontal load. The two main loads for the walls constructed as retaining structures are the loads experienced from retaining earth or water.

3. LOADS ON STRUCTURE

Loads acting on any structures are classified into various categories. The intensity of load for various types is standard and specified by the codes. However under certain conditions it would be better to ascertain the actual load to be experienced by the structure for better accuracy.

The following are the various types of loads that are considered:

3.1 Dead Load

This is also called gravity load. It may be defined as the static forces caused by weight of every element within the structure. The forces resulting in dead load consists of the weight of the load-bearing element of the building, floor and ceiling finishes, permanent partition walls, façade cladding, walls, storage tanks etc. the combined weight of all these make up the dead load. The magnitude of dead load shall comply with IS875 Part I for dead loads for various materials.

3.2 Live Load

Live loads are caused due to movements of various objects. Live loads are variable and unpredictable. Live loads can result from the occupants or by the movable objects like equipments, machines, movable partitions etc. The magnitude of live load shall comply with IS875 Part II for live loads. For a residential building the live load is taken as 2 kN/m^2.

3.3 Wind Load

The lateral loading due to wind is a major factor that causes the design of tall structure to differ from that of normal structures. Wind becomes a prominent force for buildings built near the coast. Generally above the height of 10 meters, the wind load for the building should be considered critical. Wind is dynamic in nature i.e. its nature varies with time and location, it is difficult to estimate the wind through dynamic analysis hence it is converted into equivalent static load and then the structure is evaluated and designed for these loads. The structure designed for wind load should comply with IS875 Part III for wind load.

3.4 Earthquake Load

This is another major load that is considered in the design of structures and especially tall structures. Earthquakes occur when there is a release of energy due the movements of different layers inside the earth's crust. Earthquake is dynamic in nature and predicting the time, nature, and intensity of earthquake is almost impossible. Although it is not possible to design earthquake proof, however it is possible to design the structure to resist the earthquake without significant damage. The design of earthquake resistant structure should comply with IS1893. Earthquake is measured in Richter scale. In India, the country is divided into 5 zones, zone I being the least prone to earthquake and zone V being affected by earthquake of both higher frequency and intensity.

3.5 Snow Load

Snow load occurs in places where the temperature is extremely low and the snowfall is a regular feature. Snow load needs to be considered only on the roof and other areas of the structure where snow may accumulate, such as balcony, sundeck etc. Generally the design of the building is done is such a way, so that there is limited amount of stagnation of the snow. The angle of slope is determined based on the quantity of snowfall.

3.6 Hydrostatic and Soil Pressure

The structures which are built fully or partly below the ground experience pressure exerted by soil and water. Foundation will be subjected to uplift pressure due to hydrostatic force. For structures like retaining wall, diaphragm wall, walls in basement, hydrostatic and soil pressure will act horizontally. For structures built on slopes, the columns will be subjected to soil pressure. The structures need to be designed factoring these loads.

3.7 Special Loads

The structures are subjected to internal stresses due to temperature and fatigue and other special loads. The expansion and contraction due to changes in temperature needs to be factored in design. The temperature can vary over a period of time due to climatic changes for different regions.

A structure experiences fatigue due to cyclic loading or uneven loading due to changes in shape. The structure needs to be designed for stresses experienced due to fatigue by estimating the stress cycle expected during the life of structure.

4. BUILDINGS

Building generally refers to a habitable structure. Building is generally defined as a permanent structure which serves the various needs of the society. A building acts as a shelter from weather, provides security and space for comfortable living and working.

4.1 Classification of Buildings

Based on the use, the buildings can be classified as:

(a) **Residential:** This is a type of structure which is used for living. Generally a building is defines as residential building if more than half the floor area is used for dwelling purposes.

(b) **Commercial:** This type of structure generally refers to the space being used for commercial activities such as shops and offices.

(c) **Industrial:** Industrial structures are those types of structures which are used for manufacturing. Factories are example of industrial structures. Generally industrial structures are made of pre-fabricated materials.

(d) **Institutional:** Institutional buildings are those categories of building which are used for public activities. Hospitals, Schools are some of the examples of public buildings.

(e) **Assembly Buildings:** These are those types of buildings which have large gathering of people for public activity. The activities could be for amusement, recreational, exhibition, travel etc. Examples of assembly buildings are theatres, museums, passenger terminals etc.

(f) **Mixed:** Mixed used structures are those structures which have a combination of residential and commercial use. Most of the skyscrapers are mixed use structures.

4.2 Parts of a Building

The various parts of building are shown in figure 3.6 and described as follows:

(a) **Foundation:** Foundation is that portion of the building which transfers the load of the entire building to soil and ensures stability. It is the part of the building which is below the ground or surface. Section 2.1 describes the various types of foundation.

(b) **Framework:** The framework is the superstructure on which the dead load of the roof and the imposed load is transferred. As mentioned in section 2.2, the framework can be load bearing or framed as the case may be.

(c) **Grade Level:** Grade level is the natural ground level (NGL) which separates the sub-structure and superstructure. It is also the datum which is used a reference for the heights are measured. The depth of the foundation, height of the plinth, height of the top out is all taken from NGL.

(d) **Plinth:** By definition plinth is the raised platform which supports the column. In modern terms, where the column is monolithically cast with foundation, plinth is height to which the ground floor is raised so that the building is protected from water flowing inside. On an average the plinth height is taken as 1.2 m from the natural ground level.

(e) **Basement:** The space between the grade level and plinth level is commonly referred to as basement. The basement is generally a partial sub-structure. Unless the basement is used like a car park or for any other purpose, the basement is filled with soil and other filler materials that are locally available.

(f) **Openings:** Openings are the spaces provided in the building for either the movement of human traffic or natural elements like sunlight, ventilation etc. The openings are created for doors, windows and ventilators. The openings have two portions namely:

Fig. 3.6: Figure showing parts of a building

(i) **Sill:** Sill is the flat piece which forms the base of the door or a window. In a window opening, the sill can be used as a platform for decoration too. Normally the height of sill from the floor or plinth level is 900 mm. One of the main functions of the sill is to shed the rainwater which runs down from the windows.

(ii) **Lintel:** Lintel is the horizontal member which is placed above the window or the door to support the walls above the joineries. Lintel is designed as a bending member. Commonly the height of lintel is 2100 mm. The lintel is designed to carry the triangular portion of the load above the opening. Lintel should not deflect under loading which can affect the joineries.

(g) **Roof:** Roof is the top covering of building. Roof is supported either, on walls in a load bearing structure or monolithically built with beams and connected to columns in framed structure. Roof can be flat roof or sloped roof depending on the design requirements.

In earlier days materials like thatch and clay tiles were used as roof covering. These materials were fragile and needed to be maintained regularly. With the introduction of concrete, a more permanent solution came into being.

(h) **Parapet:** Parapet is a low protective wall along the edge of the roof or a balcony or a walkway. The word is adopted from the Italian word parapetto meaning to cover. Traditionally this is the upper most construction done on the terrace. Generally the height of the parapet is taken as 1.2 m. Normally, wall of half thickness is constructed as parapet but present day parapet is made of various materials like cast iron, stainless steel or glass.

Reference and Further Reads

Prakash, Shamsher, Hari D. Sharma, "Pile Foundation in Engineering Practice", 1990, Wiley Publication.

Punmia, B.C., Ashok Kumar Jain, Arun Kumar Jain, "Soil Mechanics and Foundations", 16th Edition, 2005, Laxmi Publication IS875—Parts I to V.

Tomlinson, M.J., "Foundation Design and Construction", 7th Edition, 2001, Prentice Hall.

Varghese, P.C., "Foundation Engineering", 2012, PHI Learning Private Limited.

4 CONSTRUCTION PROCESS

The construction process involves various activities starting from the production of the raw material, selection of the equipments to be used, estimating the work to be done to the actual process of the building a structure. Though the process of construction has remained more or less same over the years, the invention of new materials, equipments and techniques, both in design and process, have had major impact in present day construction.

The design of a structure is done from top to bottom whereas the construction is carried out the other way around. The basic process remains the same irrespective of the quantity. Each activity has its own importance in the sequence of a construction process. The sequence of activity is very important and missing out on the sequence has long-term effects on both the progress of construction and the life of a structure, especially in a RCC structure.

1. PRODUCTION OF CONCRETE

Concrete is the most popular material used in present day construction and hence the process of production concrete has to be studied in detail. The production of concrete has the following steps:

1.1 Storing of Material

The storing of material is a very important process in construction. Improperly stored material can affect the properties of the materials which in turn will affect the quality of finished material. Also the materials have to be stored in such a manner that the materials don't mix before the batching is done. The various material to be used in the production of concrete have set pattern of storing.

(a) **Cement:** Cement manufactured using chemical substances especially oxides can start chemical reaction when exposed to atmosphere. Hence IS code stipulates that the cement be stored in a tightly covered chamber where it is not exposed to moisture from all sides. Furthermore, the cement bags should not be stacked to a height of more than 10 bags in each stack.

(b) **Aggregates:** Aggregates should be stored in open bins near the batching plant or the production site. Where there are no bins provided, the aggregates should be stored on a clean, flat surface and placed in a shape of a trapezoid.

(c) **Water:** Water should be stored in a hermetically sealed tank. It should be ensured that water is not contaminated by unwanted materials like chemicals. The tank should be cleaned periodically.

1.2 Batching

Batching of concrete is defined as that process where in the materials used for production of concrete is measured for the required ratio. IS456 clearly states that the proportion of cement, aggregate and water should be measured by weight. However for smaller quantities, measurement is done by volume too taking into account the corrections for the ratio of weight and volume. Hence on site there are two methods of batching adopted namely:

(a) **Volume Batching:** As the name suggests, the measurement is done on the basis of volume. Though it is an approximate method, in India for majority of smaller construction projects this method is widely adopted. This batching is suggested only when the grade of concrete is limited to not more than M20.

Most common way of measuring is by pans. Normally the quantity is measured by the number of pans used to fill the volume equal to one bag of cement. The volume of cement is almost 1.25 cft or 0.035 cum. The quantity of sand is increased for compensating for the bulkage of sand. For an M20 grade concrete, the measurements of material is taken as 2 pans of cement, 3 pans of coarse aggregates, 7 pans of sand and water is added as required for workability.

For a more accurate batching the quantity of materials is measured using a box. The size of the box is (300 × 300 × 380) mm which is equal to the volume of one bag of cement.

(b) **Weight Batching:** This is the appropriate method of batching. According to IS456, the proportions of ingredients should be taken only by the ratio of their weight. For higher grade of concretes, where the mix is designed, weight batching is the only method of batching to be used.

1.3 Mixing

Mixing is technically the first process in the production of concrete. Mixing has to be done in a proper way so that the concrete produced is homogenous. There are two methods of mixing namely:

(a) **Hand Mixing:** As the name suggests, in this process the mixing is done manually. In earlier days, most of the concrete produced in India was done using this process. Following are the steps adopted in this process:

- A flat surface is prepared which is clean and void of any external particles. Generally a steel plate of size 2 mt × 2 mt is taken.
- Fine aggregate is measured and placed on the surface in the form of a mound and cement is mixed dry. The mix should have a uniform colour and should be homogenous.
- The coarse aggregates are added and again mixed dry. Shovel is used for mixing. Subsequently required amount of water is poured for workability and the mixed to obtain concrete.

(b) **Machine Mixing:** For larger quantities and for a faster production of concrete, concrete mixers are used to produce concrete. For smaller quantities, portable concrete mixer is used. The capacity of the mini mixer is equal to concrete produced by 1 or 2 bags of cement. For larger quantities, the mixing is carried out in batching plants.

1.4 Transporting

Transporting is the process of carrying the fresh concrete mix from the point of batching to point of placing. The concrete should be transported before the concrete starts setting. Generally the initial setting time of concrete is 30 minutes, so ideally the transporting time of concrete is less than 30 minutes. However where there is a possibility of delay of transporting beyond the 30 minutes, admixtures namely retarders are mixed to ensure that there is a delay in the initial setting of concrete. The type of equipment is used for transporting depends on the quantity, nature of work, distance between the point of mixing and point of placing. The following are the equipments used to transport concrete:

(a) **Pans:** Using pans to transport the fresh concrete is the basic method of transportation which has been followed from the early days. The method is used now-a-day when the concrete quantity is small and the distance to be covered is less. The disadvantage of this method is that it is time consuming and expensive.

(b) **Wheel Barrow:** Wheel barrow is used for transporting of concrete, when the concrete quantity is larger and the surface is flat. This type of transporting is generally limited to concreting for foundation where there is not much hindrance for movement of wheel barrow.

(c) **Mini Lift (Rope and Bucket):** This type of system is used when concrete has to be transported to a reasonable height and quantity is relatively less. It is generally adopted up to a height of 2 – 3 storeys. The maximum quantity which can be transported in a single lift is generally up to 2 tons.

(d) **Builder's Hoist:** This type of system is used when concrete has to be transported to a reasonable height and quantity is relatively less. The maximum quantity which can be transported in a single lift is generally up to 5 tons. The builder's hoist consists of a platform with a set of steel frames erected to the required height. The system can be used to transport not just fresh concrete but also the raw materials required for construction.

(e) **Belt Conveyor:** Belt conveyors are generally used for longer distances on a horizontal place where the access is limited. This type of system is generally used for concreting in tunnels where the concrete has to be transported quickly over long distance.

(f) **Tower Crane:** Modern technology has enabled concreting for longer and higher distances and made construction process much faster. Tower cranes are used for large construction projects where the distance to be covered is large.

1.5 Placing

The next process after the transporting of fresh concreting is pouring or placing of concrete on the formwork. For RCC works, the concrete should not be poured in a layer of more than 150 mm to 300 mm in a single pour. However in case of mass concreting, the depth of layer can be up to 450 mm. Following are the precautions to be taken during the placing of concrete:

- Formwork should be properly placed in required shape and be properly secured.
- The surface of the form should be well cleaned and well oiled so that the hardened concrete does not stick to the form and it is easy for de-shuttering.
- The critical height of pouring should not exceed 1.5mt to prevent segregation of concrete.
- Concrete placed should be thoroughly worked around the reinforcements.
- The latence formed in any layer should be removed and the surface should be roughened before placing of next layer.
- As far as possible the concrete should be placed monolithically and joints should be avoided. However in case of formation of cold joints, proper procedure for subsequent concreting should be followed after the treatment of the joint.

1.6 Compacting

Compaction is a very important process in the production of concrete structure. It ensures that concrete is homogeneous and achieves the desired strength it is designed for. Compaction has following purpose:

- It removes air entrainment and fills up the voids with solid substance.
- It ensures the integrity of concrete and creates homogeneity.
- It ensures a higher density of concrete.

Generally an excess of 5% of voids in concrete due to improper can reduce the strength of concrete by almost 30% and a 10% void reduces the strength by almost 50%. Compaction is done by two methods namely:

(a) **By Hand:** In this method the compaction is done by hand using either a rod or a flat. Generally the rod is used for beams and columns and is done using a mild steel rod of 16 mm in diameter and 2 mts long. The flat is used for slab concreting and it is called tamping. The flat is generally a square section made of either wood or aluminum. Compaction by hand has almost been done away with because it is time consuming and laborious.

(b) **By Mechanical Vibrators:** A mechanical vibrator consists of a rotating needle having a diameter ranging from 20 mm to 90 mm and a length ranging from 300 mm to 650 mm and connected to the base which holds the motor. Generally the motor is run on liquid fuel. The frequency at which the needle vibrates is in a range of 3000 to 6000 rpm. The specification of a mechanical vibrator is described in detail in IS1161.

The different types of vibrators used are:

(i) **Internal Vibrator:** This is the most commonly used vibrator on construction site. This type of vibrator is immersed in fresh concrete and the process of vibration takes place. The immersion should be for few seconds (30 secs to 100 secs) only and the needle should be immersed vertically. Improper vibration will result in segregation of concrete or voids in concrete which will directly affect the strength of the concrete being placed.

(ii) **Formwork Vibrators:** These are external vibrators which are clamped on to the formwork at predetermined points. Here both the form and the concrete are vibrated. These types of vibrators are generally used in thin precast sections where the needles cannot be immersed.

(iii) **Surface Vibrators:** These are used for compacting shallow surfaces. Generally this type of vibrator is used for pavements and compaction of sand base for laying of paving blocks.

(iv) **Table Vibrators:** These vibrators are generally used for compacting of precast blocks. These vibrators consist of platform on which the sections are place and then vibration is carried out.

1.7 Curing

Curing is the most important process in production of concrete which ensures the strength of concrete placed is achieved and retained. Cement concrete dissipates heat of hydration due to the chemical reaction which takes places in the formation of concrete. Due to the evolution of heat of hydration, the water present in the concrete will evaporate into atmosphere. Water is a very important ingredient in concrete and without water the chemical reaction will be incomplete. Hence it is important to ensure that the water does not evaporate into atmosphere. Curing is the process of ensuring that the water required for the chemical reaction for producing concrete is kept intact. The different methods of curing are as follows:

(a) **Moist Curing:** In this process, the surface of concrete is kept moist with the help of water. For flat surfaces, like slabs and pavement, the surface is kept moist by creating water ponds and maintaining the required quantity of water for the required curing period. Cement bunds are created all around the surface to a height of 75 mm and the water is stored on the surface like a pond. In vertical surfaces, like columns and retaining walls, jute cloth or hessian is wrapped around the surface and the jute cloth is constantly sprayed with water to ensure that the jute bag is kept moist at all times for the entire duration of curing period. For other exposed surfaces, water is sprayed periodically to keep the surface moist and ensure that proper curing is carried out. This is the most efficient method of curing, though this type of curing is a bit laborious and requires constant supervision.

(b) **Membrane Curing:** This method is generally adopted in areas where this is shortage of water. A membrane is used to cover the exposed surface of the concreted element to ensure moisture is retained. The membrane could be either in solid or liquid form. Though this process does not require supervision after the membrane is laid, it efficiency is lower than moist curing.

(c) **Steam Curing:** This type of curing is done in cold weather condition. In cold weather, the water contained concrete tends to solidify and hence it is necessary to ensure the water is maintained in liquid form. To ensure this, steam is sprayed on the concrete surface. The temperature of steam is dependent on the atmospheric temperature.

1.8 Finishing

Once the concrete is set, the formwork has to be removed within a stipulated time so that the formwork doesn't stick to the concrete and there is no damage to the concrete member while removal. IS456 stipulates the standard time for the removal of formwork for various members.

When the formwork is removed, the surface finish may not always be smooth and pleasing to the eye. Alternatively, where the quality control is low, there may be visible voids. These voids are generally known as honeycombs. Any voids or openings which allow the moisture to percolate in the concrete will not be good for durability and strength of structure. Percolation of moisture into the concrete will expose the steel reinforcement and facilitate the process of oxidation of steel. Oxidation of steel will cause the steel to rust and hence affect the strength and durability of concrete. Hence it is necessary that the concrete surface is finished properly. The various types of finishes adopted are:

(a) **Formwork Finish:** In mega structures and on surfaces where architectural finishes are not required, it is ensured that the quality of formwork is maintained very high. Generally in these areas, the material used for concrete is such that after de-shuttering, the finish is proper and as per desired standards. This type of finish is adopted for structures like bridges, in the ceiling of floors of building which are used for car park and other places where plastering is to be avoided. Generally the type of formwork adopted for this type of finish is plywood. However modern technology has facilitated the use of more durable materials like aluminum and fibre. These types of formwork are expensive and capital intensive and used only in large construction projects.

(b) **Applied Finish:** This type of finish generally indicated the application of extra finishing after the formwork is removed. The most common type of applied finish is plastering with cement mortar. This type of finish is adopted where the quality control on finishing is slightly lower. Here the cost of formwork is lesser and the type of formwork used is MS folded plate for slabs and some locally used wood for sides of beams and slab.

2. FORMWORK

Formwork is the term given to the mould over which the concrete is poured. Formwork is generally a temporary structure and is removed after a certain period of time depending on the type of work and its requirement. It is also called shuttering. Formwork is not only used for supporting the pour of concrete but also used in supporting brickwork to form arches and other shapes.

2.1 Requirements of Good Formwork

A good formwork should satisfy the following:

- It should be strong enough to withstand all types of dead and live loads.
- The material of the formwork should be cheap, easily available and should be suitable for reuse.
- It should be water proof and should not get affected due to the moisture present in concrete.
- The material of the formwork should not warp or get distorted when exposed to the elements.
- It should have a smooth surface and give a good finish to the concrete after the formwork is removed.
- It should be rigidly constructed and efficiently propped and braced both horizontally and vertically, so as to retain its shape.

- The joints in the formwork should be tight against leakage of cement grout.
- Construction of formwork should permit removal of various parts in desired sequences without damage to the concrete.
- The formwork should be set accurately to the desired line and levels should have plane surface.
- It should be as light as possible.
- It should rest on firm base.

2.2 Loads on Formwork

The formwork has to withstand various types of load and remain stable for the duration of its use. Hence the formwork and its support have to be designed to carry these loads. The various types of load on formwork are as below:

- Self weight of formwork.
- Live load during construction.
- Dead load of wet concrete.
- Hydrostatic pressure of fluid in concrete.
- Impact load due to pouring of concrete.

2.3 Types of Material for Formwork

The various types of material used for formwork are as below:

(a) **Timber Formwork:** This is the cheapest material used for formwork. Timber for formwork should satisfy the following requirement:

- It should be well seasoned and not green.
- It should be light in weight.
- It should be easily workable with nails without splitting.
- It should be free from loose knots.

The traditional timber formwork consists of supports made out of lumber or young tree trunks, which support rows of stringers, assembled 1 to 2 metres apart, depending on thickness of slab. Between these stringers, joists are positioned roughly 30 centimeters apart upon which boards are placed. The stringers and joists are usually (100 × 100) mm or (100 × 150) mm lumber. The most common wood thickness is 18 mm.

(b) **Plywood Formwork:** Plywood formwork is used in places where the finish is required to be good. Generally in case of fair face finish or exposed concrete works, plywood is used. Resin bonded plywood sheets are attached to timber frames to make up panels of required sizes. Though the cost of plywood formwork expensive compared to that of timber shuttering and it may prove cheaper in certain cases in view of the following considerations:

- It is possible to have smooth finish in which case on cost in surface finishing is there.
- By use of large size panels it is possible to effect saving in the labour cost of fixing and dismantling.
- Numbers of reuses are more as compared to timber shuttering. The number of reuses can be taken as 20 to 25.

(c) **Steel Formwork:** This is the most widely used material formwork especially in India. The steel formworks are manufactured in form of folded plates for slabs as well as beam side form. They are supported on steel joist which in turn is supported on steel pipe sections (more popularly known as jacks). In present day construction, forms made of steel folded plate are most common material used for formwork in case of slab. The biggest advantage is the number of reuses which makes it cost affective. The disadvantage is the unevenness in the finish of concrete. It may not be possible to get the same finish as plywood shuttering and it cannot be used in exposed concrete works.

(d) **Aluminum Formwork:** Aluminum formwork is a comparatively new technology. The biggest advantage is re-use cycle which makes it cost effective. There are materials which can be reused up to 300 cycles. The other advantage is that the finish is smooth.

(e) **High Density Plastic Formwork:** Plastic formworks are becoming popular because of its light weight, and good finish requiring little surface treatment. Because of its flexible nature, the formwork can be moulded to any shape. This type of mould is extensively used in waffle slabs.

2.4 Types of Formwork

There are different types of formwork available for different purposes. Generally, the formworks for vertical concreting are called wall forms and those for horizontal concreting are called slab or floor forms. The various types of formwork available today in the market are as below:

(a) **Traditional Formwork:** The traditional slab formwork technique is similar to a framed structure. The system consists of horizontal panels supported by joists which in turn are supported on vertical column like support sections. This is the simplest type of formwork and widely used in construction of buildings.

(b) **Table Formwork or Flying Form:** These systems consist of slab formwork "tables" that are reused on multiple stories of a building without being dismantled. The assembled sections are either lifted per elevator or "flown" by crane from one story to the next. Once in position the gaps between the tables or table and wall are filled with "fillers". They vary in shape and size as well as their building material. The use of these systems can greatly reduce the time and manual labor involved in setting and striking the formwork. Their advantages are best utilized by large area and simple structures.

(c) **Climbing Formwork:** Climbing formwork is a special type of formwork for vertical concrete structures that rises with the building process. While relatively complicated and costly, it can be an effective solution for buildings that are either very repetitive in form (such as towers or skyscrapers) or that require a seamless wall structure (using gliding formwork, a special type of climbing formwork).

(d) **Sliding or Slip Formwork:** Slip form is type of formwork where the formwork moves continuously with such a speed that concrete when exposed has already achieved enough strength to support the vertical pressure from concrete still in the form as well as to withstand lateral pressure caused by wind etc.

Thus, the slip form concreting technique is a rapid and economical construction method that can be applied with great advantage to many types of construction projects such as chimneys, silos, water towers, bridge-columns, lift shaft cores and shaft lining etc.

The technique is based on movable forms which are gradually lifted by hydraulic jacks. It is a continuous process where wet concrete is added to wet concrete. Reinforcing steel and/or post tensioned cables are continuously fixed as the normal slipping speed is 3 to 6 meters per 24 hours. The slip form construction is designed for each project depending on the structure of the project. The advantages of slip forming are:

- Minimum consumption of timber and steel plates.
- Total elimination of traditional scaffolding.
- Minimum requirements of carpenters for assembling.
- It gives a monolithic structure.
- The concrete surfaces can be treated and finished while concrete is green.
- Depending on the weather conditions, it is possible to achieve a vertical rise to the tune of 4 to 5 m in summer and 2 to 3 m in winter.

(e) **Tunnel Formwork:** The tunnel formwork is a room sized structural steel fabricated form which is used to cast the RCC walls and floor slabs of a building as a monolithic structure in a continuous pour. The forms are then heated using hot air blowers for accelerated curing of the concrete. This system is most economical when the structure consists of large number of identical units. There exist two versions of this type of formwork namely:

- The half tunnel formwork used to cast only one wall and slab simultaneously.
- The full tunnel formwork used to cast two walls and a slab simultaneously.

The sequence of construction involves placing of reinforcement, electrical and sanitary conduits along with the tunnel forms. Concrete is then poured and the open side of the forms is covered and hot air blowers placed inside. The forms are removed the next day and placed on the next site using cranes. The optimum use of tunnel form is in multiunit shear wall structure with identical floor layout at each level.

2.5 Removal of Formwork

The removal of formwork is also called stripping or striking of formwork or de-shuttering. The removal of formwork needs to be carried out carefully and as per specification. If the removal is carried out earlier than the time specified, then the concrete will disintegrate. If the formwork is left too late, then the formwork will set with concrete and removal will damage the concrete surface. The duration after which the formwork can be removed is called the stripping time. According to IS456, the duration after which the formwork can be removed is shown in Table 4.1.

Table 4.1: Table showing the stripping time of formwork

S.No.	Type of Formwork	Minimum Period
1	Vertical formwork to columns, beams and slabs	16-24 hrs
2	Soffit Formwork to slab (Props to be refixed immediately after removal of formwork)	3 days
3	Soffit Formwork to beams (Props to be refixed immediately after removal of formwork)	7 days
4	Props to slab spanning up to 4.5 mts	7 days
5	Props to slab spanning above 4.5 mts	14 days
6	Props to slab spanning up to 6.0 mts	14 days
7	Props to slab spanning above 6.0 mts	21 days

3. STEPS INVOLVED IN CONSTRUCTION

The sequence of work in construction is a very important factor in executing a project. Any variation in the sequence will result in cascading effect in terms of time, quality and cost. The construction of a building has the following activities or steps in the sequence:

3.1 Site Preparation

Before the construction is started, the site where the construction is to be carried out has to be prepared. The site should be leveled and any obstructions should be removed. There should be no trees with deep expanding roots near in the vicinity of foundation of the structure. The boundaries have to be marked properly and a preliminary survey of the outer boundary has to be carried out. The places for storage of material, labour sheds and the path of movements have to the clearly earmarked. The storage of material should be done such that there is no hindrance to removal of other materials and restocking of material and to the movement within the site.

3.2 Setting Out

Setting out is the first activity carried out on site. For a smaller size project, generally the setting out is done using the conventional methods. The measurement is done using tape and the markings are pinned using steel rods. The rods are then connected using strings. Right angles are checked before the final marking is fixed. The measurements are then transferred to a more permanent position either on a nearby wall or bed blocks constructed using either bricks or blocks. The most common technique to check the right angle on site is the use of 3-4-5 method or Pythagoras formula.

For larger projects and for projects where the stakeholder desires accuracy, the setting out is done using mechanical means namely Total station. Total station gives an accurate position of all desired points.

3.3 Excavation

This is the first activity carried out after the setting out of the centerline has been done. Depending on the quantity of excavation the process is carried out either mechanically or manually.

Fig. 4.1(a): Figure showing site preparation

The excavated earth is stored at a nearby location so that the earth is used for backfilling once the foundation is cast and column is erected above the ground level. The depth of excavation is based on the design requirements.

Fig. 4.1(b): Figure showing setting out

Fig. 4.1(c): Figure showing excavation

Fig. 4.1: Figure showing setting out and excavation process

3.4 Sand Filling

After the excavation is carried out, the surface is filled with sand up to a maximum of 150 mm. This is to ensure that the base is uniform. This is done only for shallow foundation. Sometimes when the subsoil is weak, like in black cotton soil or weak clay, then a process known as soil stabilization is carried out to increase the bearing capacity of the sub-grade.

3.5 Pest Control

Pest control is carried out to ensure that the concrete is protected from the infestation of pest. This is also referred to as anti-termite treatment. Pest control is carried out in three stages, namely below the foundation, below the ground floor and in the joint of floor and walls. This ensures that the termites don't enter the building and the components of building especially the wooden elements are protected.

(a) **Pest Control Below Foundation:** Pest control treatment or anti-termite treatment is done just before laying of PCC and after the filling of sand. Generally the pest control is in form of liquid and it is done to ensure that it covers the sand fill homogenously. This process should be carried out in dry climate and the sad should not be saturated. Laying of PCC should be done immediately on the completion of the process. This is done to ensure that the effect of pest control doesn't evaporate into atmosphere and render it ineffective.

(b) **Pest Control at Plinth Level:** This process is the same as the previous process. It is done after the sand filling in basement at plinth level.

(c) **Pest Control After the Construction of Walls:** This process is carried out after the walls are constructed and before plastering. In this process, the solution of anti-termite treatment is pumped into the edge of the wall and floor.

3.6 Laying of PCC

Once the soft bed of sub-grade is completed, the sub-grade needs to be given a hard platform to carry out the next step, namely, laying of foundation. The hard bed is made using Plain cement concrete. The reason it is called Plain Cement Concrete or in short PCC is because it is not reinforced and it is not used to carry any structural load. Its only use is to provide a flat bed for the structure and it provides extra concrete cover from the sub-soil. Generally the PCC is made using larger sizes of aggregates. The sizes of aggregates can go up to a maximum of 63 mm. The grade of concrete used is M10 to M15.

3.7 Casting of Foundation

Casting of foundation is the first step in the process of actual construction. Foundation is the structure which will withstand the load of the entire structure and hence it is very important that the casting of foundation is done with utmost care. Furthermore, being a substructure and below the ground, rehabilitating the foundation after construction becomes a very difficult process. Casting of any RCC member will have the four processes, namely placing of formwork, placing of reinforcement, pouring of concrete and removal of formwork. However in case of foundation and raising of columns, the formwork is placed after the placing of reinforcement as the only formwork used for foundation and columns is the side form.

(a) **Placing of Reinforcement:** Placing of reinforcement is the first step in the process of casting of foundation. Reinforcements are cut and bent as per design requirement and placed on the PCC. Cover blocks are placed below the foundation as per requirements. For foundation works, IS456 specifies that the clear cover to main reinforcement should be 50 mm.

(b) **Erection of Column Reinforcement:** After the reinforcement for foundation is placed, the reinforcement for column is placed on the foundation reinforcements. Generally the height of the column reinforcement is taken to a maximum of 6 mts which generally works out to just below the roof of ground floor.

(c) **Placing of Side Form:** For foundation, the only formwork placed is the side form at the edge of concrete line. Generally the side form used is of timber which can be removed after the setting of concrete. However sometimes the removal of formwork becomes cumbersome and under such cases bricks or blocks are used as side form. After the completion of concrete, the blocks are left to remain in the foundation. Sometimes the side form is avoided and the excavation is done in such a way that the edge of excavation forms the edge of concreting. This can be adopted in case the sub-soil is hard like rock where there is no possibility of soil sliding during concreting. The disadvantage of such a process is that the side finish may not be uniform and there will be a variance in quantity of concrete.

(d) **Pouring of Concrete:** This activity is carried out after the placing of reinforcement and formwork. Before pouring of concrete it is to be ensured that the place where concrete is being poured is clean and devoid of any external objects. For isolated footing, the concreting is self-compacting as it is poured from a height and there is no need for vibrating. However in case of all the other types of footing it is necessary to carry out proper vibration. While concreting of foundation care has to be taken to ensure that the column reinforcements are held and secured in correct position.

(e) **Removal of Formwork:** The removal of formwork should be done after the concrete is set and it is generally 24 hrs after the pouring of concrete. The removal of concrete should be done taking into account the structural design, e.g., For cantilevered portion, the removal of formwork should commence from the free end. While removing formwork, care should be taken so that the concrete is not chipped. It is also to be ensured that no part of formwork is retained in the concrete.

(f) **Curing:** It is the most important activity after removal of formwork. For foundation, generally the curing is done by spraying of water or placing hessian cloth and keeping the cloth moist all the time. Curing should be carried out for 7 days.

3.8 Raising of Column to Grade Level

After the foundation is completed, the next process is to raise the columns up to grade level.

(a) **Casting of Starter Block:** After the completion of foundation concreting, the starter block for column is cast. Generally the height of the starter is 10 mm and the concrete mix is the same as the one to be used for column. Before casting the starter, the setting out should be checked for centerline of columns. Started block also helps in maintaining the formwork in position.

(b) **Erection of Reinforcements:** Generally the main reinforcements are already in place during the casting of foundation. In this stage the ties rods are placed is necessary and the main reinforcements are secured. The cover blocks for the main reinforcements are tied to the main reinforcement.

Fig. 4.2(a): Raising of column upto plinth level

Fig. 4.2(b): Grade beam

Fig. 4.2(c): Starter block

Fig. 4.2(d): Raising of column above plinth level

Fig. 4.2: Figure showing raising of column

(c) **Placing of Side Form:** The side forms are placed and the verticality is checked with the help of plumb bob. Before placing the formwork should be oiled properly so that the concrete doesn't stick to the formwork and the removal of formwork is easy and the concrete surface is not damaged during the removal of formwork.

(d) **Pouring of Concreting:** Pouring of concrete is generally done manually for column concrete. Since the concrete is poured from a height the compaction is partially done using tamping rods. Mechanical vibrators are generally avoided for smaller size columns as the forms can split due to vibration.

(e) **Removal of Formwork:** The removal of formwork is done after 24 hours as is the case for all the vertical sections. After removal of formwork the surface is checked for any irregularities such as honeycombs. The rehabilitation is carried out by adopting proper procedures.

(f) **Back Filling:** Back filling is the process of refilling the remaining portion of excavation which had not been concreted. Generally back filling is done by the same soil which has been excavated. This is done to ensure that the soil properties are uniform below the ground level. Sometimes the excavated soil is insufficient to complete the fill. Under such circumstances, additional soil is procured to fill up the excavation. The backfilling needs to be compacted properly.

(g) **Curing:** Curing of Columns is done by tying a hessian cloth around the column and keeping the column moist continuously by spraying water at regular intervals. The curing is carried out for 7 days.

3.9 Laying of Grade Beam

The steps for laying of grade beam is somewhat similar to casting of foundation.

- A firm base is created using sand base and then the PCC.
- The reinforcement is placed as per design requirement and the side forms are placed.
- The pouring of concrete can be done manually or by pump concrete if the quantity is large.
- The vibration is carried out using mechanical vibrators.
- Stripping of side form is done after 24 hours as the formwork is vertical.
- After stripping of side form, if any undulations or honeycombs are observed, then the repairs is carried out as per the procedure specified by the engineer in charge.
- The curing is carried out by spraying of water at regular interval and ensuring that the surface of the beam is always moist.

3.10 Raising the Columns to Roof Level

The procedure of raising the columns to roof level is the same as the one which was adopted to raise the columns to grade level.

- The reinforcements are erected as per design. If the column is being raised above the roof level, then the main reinforcements need to protrude above the roof for the required height.

- The lapping should be done as per design and it should be ensured that the laps of the reinforcements are staggered.
- The spacing of ties should be checked.
- Cover blocks are to the tied to the main reinforcements.
- The starter block should be cast above the plinth after checking for the setting out of the column positions.
- The side form should be placed after proper preparation and the verticality is to be checked. It should also be ensured that the verticality is not affected while concreting. The side forms should be properly secured using props.
- Placing of concrete is to be done after making adequate provision for pouring of concrete.
- The concrete should be placed in layers of 1 mt and compacted using a tamping rod of 16 mm dia MS rod.
- Stripping of side form is done after 24 hours as the formwork is vertical.
- After stripping of side form, if any undulations or honeycombs are observed, then the repairs is carried out as per the procedure specified by the engineer in charge.
- The curing is carried out by tying and hessian cloth around the column and spraying of water at regular interval and ensuring that the surface of the column is always moist.

3.11 Raising the Plinth

Once the columns are raised, the next step is raising the plinth. This will also ensure that there is a firm base for placing of props for formwork. The steps involved in raising of plinth are as follows:

(a) **Construction of Wall:** The first step in raising of plinth is the placing of block work. The level to which the plinth needs to be raised as per drawing is marked on the column. The block work is placed on the grade beam. It also acts as a retaining wall for the soil to be filled.

(b) **Filling of Basement:** Commonly the space between the grade level and the plinth level is referred as basement. After the wall is raised, the gap is filled with soil to the required height.

Generally the filling is done with the excavated soil from the site. However in most cases the quantity is insufficient. The excess soil required is procured. The most common material for filling is pit sand which is a mixture of excavated soil. When filling sand or natural soil is not available, alternative filling materials are used. Generally fly ash is used as an alternative filling material. Care should be taken to ensure that the compaction is proper and there is no settlement of floor once the floor of the structure is loaded. Compaction should be carried out till such time it is confirmed that the filled up soil has stopped settling.

(c) **Sand Filling and Laying of PCC:** This procedure is similar to the one carried out as done below foundation and plinth beam. The laying of PCC before the casting of roof ensures that there is a firm base for the props used for formwork of roof. The grade of concrete can be M10 or M15 depending on the site condition. The aggregates used for this type of PCC are generally larger, i.e. 40 mm. This will give the surface better resistance to abrasion. The thickness of PCC is generally 150 mm.

3.12 Casting of Roof

The steps in casting of roof are:

(a) **Placing of Formwork:** This is the first step carried out in casting of roof. The sequence to be carried out in placing of formwork is as follows:

- The level of soffit of beams and slabs are marked. Generally the column is raised to the point of soffit of beam.
- The formwork for the soffit of beam is placed.
- The side form for the beam is placed. The height of side form should be such that the top of the side form is the level of the soffit of the slab.
- Alternatively the formwork of slab is placed on the top of the formwork of the beam sides.
- The formwork has to be properly oiled so that the concrete does not stick to the formwork.

Fig. 4.3(a): Reinforcement

Fig. 4.3(b): Mixing

Fig. 4.3(c): Pouring concrete

Fig. 4.3(d): Vibrating

Fig. 4.3: Figure showing various casting of roof

(b) **Placing of Reinforcement:** Once the formwork is placed the reinforcements are place in position. First the reinforcements for the beams are placed and then the reinforcements for slab is placed anchored to the beam for bearing. Care should be taken to ensure that the reinforcements are devoid of any rust.

(c) **Leveling of Formwork:** Once the reinforcements are placed in position, the level of the formwork has to be checked and any adjustments required are to be made. Following are the points which are needed to be ensured:

- It is to be ensured that the levels are as per design requirements.

- There are no gaps between the joints of the formwork. Gaps between formwork joints will allow the concrete slurry to seep and thus affecting the concrete mixture.
- The props are adequately braced and secured at the base. The bracing should ensure that the formwork is secure during the placing of concrete.

(d) **Placing of Concrete:** The concrete is poured after it is conferment by the engineer in charge that the formwork and reinforcements are ready in all respects. Concrete can be placed either manually or by pump concrete depending on the quantity and site condition. The procedure and standards are the same as mentioned in production of concrete (item 2). The concreting should take of the following steps:

- The concreting of the beams adjacent to the slabs on both sides should be carried out first.
- This is done to ensure that the differential loading due to placing of concrete and thermal effects is taken care of.
- Vibration needs to carried out properly so that concrete flows into every nook and corner and voids are minimized.
- Slab concreting is to be carried out and the compaction is to be done by tapping.
- Spreading of concrete using needle vibrator for slab should be avoided. This will result in segregation of concrete.

(e) **Curing:** The process of curing should begin immediately on the setting of concrete. The setting of concrete here refers to final setting time which is generally 600 mins. However with higher grade of concrete, the setting can take place earlier. The engineer in charge should visually observe the surface of the concrete before the process of the curing is commenced. Any micro cracks observed should be immediately filled with cement slurry. The curing for slab is carried out by process known as bunding which is shown in figure 4.3. The duration of curing should be carried out as specified in IS456.

(f) **Removal of Formwork:** The removal of formwork is also known as de-shuttering. The duration after which the formwork can be removed is specified in IS456. Generally for normal works it is done after 14 days from the date of completion of casting. Under special circumstances the de-shuttering can be carried out earlier provided adequate provisions are made. The time of de-shuttering is specified in Table 4.1.

3.13 Construction of Wall

The construction of wall in superstructure has the following sequence:

(a) **Construction of Wall up to Sill Level:** Once the formwork for the roof is removed, the construction of wall in superstructure can commence. The first stage is to construct the wall up to sill level. Sill level is the level over which the windows are placed. Normally for any residential building the sill level is taken as 900 mm from the floor level. Care should be taken to ensure that the wall is in a straight-line. This will ensure a uniform thickness in plastering. It is easier to maintain a straight-line in a block work construction rather than brickwork as the units of blocks are longer and number of joints is lesser. In case of the inner walls, there is no sill and the wall is built to the height of lintel level or the soffit of the beam or roof as the case may be.

(b) **Casting of Sill Beam and Marking of Openings:** The brickwork can experience diagonal tension due to the thermal properties of joineries. The diagonal tension can result in diagonal crack in the wall. Though in a framed structure this will not have any impact on stability of the structure, the crack can show poor surface finish over a period of time and be unpleasant to the eyes. To avoid this, a RCC slab is cast over the wall, know as sill slab. Sill slab can be cast using nominal grade concrete to height of 100 mm and using nominal reinforcements. This will act as a bed block and ensure that the load is distributed evenly and the diagonal cracks are avoided.

Once the sill beam is cast, the openings for windows are marked as per design. The markings are done using cement slurry so that the markings are permanent and don't get erased when the wall is being raised to the next level. While marking for the openings, care should be taken to note whether markings are specified for finished surface or bare surface. If it is for finished surface, then allowance should be made for the thickness of surface finish. In earlier days and places where wooden joineries are used, the joinery is prepared after the surface is finished. The actual size at site is measured and then the joineries are made. However in today's scenario most of the finishing materials are prefabricated and hence it is important to ensure that proper space is earmarked.

(c) **Raising the Wall to Lintel Level:** After the sill beam is cast and the window openings are marked, the wall is raised to lintel level. Generally the height of lintel level is 2100 mm which essentially means the size of opening for window is 1200 mm. However depending on the architectural design, the lintel level can change.

(d) **Casting of Lintel:** Once the wall is raised to lintel level, the lintel beam is cast and placed. Lintel can be either a continuous lintel or cut lintel as per design requirements. In continuous lintel, the beam is cast for the entire length of the wall. Continuous lintel is cast *in situ*. However to cut costs and time, cut lintels are adopted. The cut lintels are lintels placed only for the opening. The size of the lintel here would be the width of the opening and the bearing on both sides of the opening to rest on wall. Lintel is designed as a beam.

(e) **Raising the Wall Above Lintel:** Once the lintel is cast, the wall is raised up to the soffit of the beam or slab whichever may be the case. It is to be ensured that there is a minute gap between the wall and the concrete surface to provide for thermal expansion.

(f) **Curing:** As mentioned earlier, curing is a very important aspect in construction. The walls constructed have to be cured at every stage to ensure that the cement mortar bonds well. Improper curing will result in surface cracks which will be visible after the wall surface has been plastered. If the wall is not cured properly, then the cement mortar will absorb water from the plastering or it might become brittle.

Fig. 4.4(a): Block work

Fig. 4.4(b): Internal plastering

Fig. 4.4(c): External plastering

Fig. 4.4(d): Painting

Fig. 4.4: Figure showing finishing works

3.14 Finishing

Finishing the structure gives it the completeness in looks. It is similar to dressing up the structure from its bare form. Finishing the structure has the following activities:

(a) **Placing of Joineries:** Once the walls are constructed, the joineries are to be placed. If wooden joineries are used then the frames for both doors and windows are fixed to the walls using clamps. The clamps are inserted on the walls and grouted. If however the joineries are made of material other than wood, then the joineries are placed after the plastering of walls.

(b) **Concealing of Service Lines:** In earlier days the service lines especially electrical lines were exposed and the electrical work was done after the completion of plastering. Due to the improvement in aesthetics and usage of tiles for cladding in toilets have made it necessary to conceal the service lines in the walls. The drawback of concealing the lines is that they cannot be changed at a later date without removal of plaster. Hence it is very important that before concealing the service lines, the interior layout has to be finalized. Concealing of electrical lines is done by chipping the walls and then embedding the PVC pipes in the walls. This is the same for plumbing lines too. The wiring is done after the plastering and painting is completed.

(c) **Plastering:** Plastering is done is two stages. Plastering is done to give the surface a smooth finish. Plastering is done using a cement mortar. Before the plastering is done, the level and thickness of the plastering is determined by placing button marks. As mentioned earlier for with cement works, curing is an important activity after plastering. Curing is done by spraying potable water over the surface at a frequency decided by the engineer in charge.

(i) **Plastering of Ceiling:** Plastering of ceiling is generally done after the removal of formwork and before the construction of walls. The ratio of cement mortar is 1:3 for concrete surface. Before plastering, the surface should be leveled by removing any protrusions of slurry. The surface should be hacked to give a rough finish for bonding. Hacking should be done before plastering for any concrete surface which is smooth.

(ii) **External Plastering of Walls:** The external plastering of wall is done once the walls, including the parapet, are constructed. Before the commencement of plastering, the scaffolding is erected for support of working platform. Sometimes the external plastering is done in 2 layers. The base coat is anywhere between 6 to 12 mm. The initial coat of plastering is done to prepare a smooth and even surface. The second layer of plaster is generally 12 mm. The second layer of plaster should be done after 24 hrs of the first layer. The cement mortar for wall plastering is in a ratio of 1:4 to 1:6.

(iii) **Internal Plastering of Walls:** Internal Plastering of wall is done after the concealing of the service lines. The cement mortar for wall plastering is in a ratio of 1:6. It is also to be ensured that, wire mesh is fixed in the joints, especially between wall and concrete and packing of service lines.

(d) **Painting:** Though painting is the final activity before the fixing of electrical plumbing fixtures, the initial coat of paint is done before the flooring is laid. Before painting is carried out, the surface is made smooth and even by applying wall putty. This ensures that any gap in workmanship in plastering is covered. Alternatively if the painting is done after the flooring is laid then it is to be ensured that that the flooring laid is protected. Generally a coat of plaster of paris is laid to protect the flooring.

(e) **Laying of Flooring:** After the first coat of painting is completed, the flooring is laid. The flooring can range from ceramic tiles which are the most popular flooring these days, to laying of expensive natural flooring as granite or marble or even manufactures timber flooring. The preparation of base for laying of flooring may vary for each of the material, in terms of the material used to the thickness of the base to match the rest of the building.

4. CONCRETING OF PILE

Concreting of piling is a complex activity and it is considered a specialization in itself. There are companies who deal only in piling activity.

4.1 Concrete Mix

The integrity of pile shaft is of paramount importance and hence the concrete mixes and the methods are directed towards this as opposed to high-strength concrete. This pre-requisite has lead to the adoption of highly workable mixes and total collapse mixes which ensure that the concrete flows between the reinforcements of pile with ease and into the interstices of soil, This type of high slump concrete is also called self compacting concrete. It is also ensured that the setting of concrete doesn't take place before the concrete is properly placed in its position inside the pile. The temperatures inside can be high and it can facilitate setting of concrete. Also the distance the concrete needs to travel can be more. To prevent the setting of concrete early, admixtures like super plasticizers are mixed to ensure the workability of concrete without affecting the strength.

4.2 Reinforcement

Generally piles are designed as compression members. However in case of hi-rise structures, the piles are also designed for lateral loading and they are adequately reinforced to resist such forces. Lateral loading can be natural forces such as earthquakes and moving loads such as movement of traffic on bridges. As it is a circular column, the minimum number of main reinforcements is 6. Unless there is a design requirement, spiral reinforcement for tie rods are avoided as it impedes the flow of concrete. The junction between the lateral reinforcements and tie should be properly secure so that the ties don't move when the cage is being lowered. Design should be such that the density of reinforcement is low for smooth flow of concrete. For piles of deeper length, there will be lapping and it should be ensure that the lapping of reinforcement is staggered as specified by IS codes.

4.3 Concreting

In the earlier days the concreting of deep bored piles used steel casings to support the soil and prevent it from sliding. However this is a tedious process as it requires heavy machinery. It also had a limitation in the depth of pile. To eliminate the use of casing the method of borehole wall support was developed from slurry trench techniques using bentonite suspension. Bentonite is a type of naturally available clay.

The process of excavation and concreting in piles is done simultaneously. This method is known as Tremie method of concreting. In this method, while excavating the bore hole is filled with a solution of water and bentonite. The bentonite solution is placed in the borehole using a pipe which is known as tremie pipe. This bentonite solution supports the bore walls by exerting hydraulic pressure on the walls. This pressure is developed by the gels formed on the wall from the bentonite solution and prevents the collapse of soil layer. The concrete is pumped through the tremie pipe and fresh concrete displaces the bentonite solution. As the borehole is filled with concrete, the bentonite solution rises. The displaced bentonite solution is collected and reused. This is reused for a cycle of 5 to 6.

5. MEASUREMENT/QUANTITY SURVEYING

Measurement is a very important part of a construction process. It is necessary to have a prior knowledge of quantity of work and material to be used and the cost involved in the project. Estimation is carried out by simple mensuration method and the cost is derived out from the quantities. The accuracy of estimation is very important it has direct implication on the finance of the project. If the finance available is less than the estimated cost of the project then ways and means are looked into to ensure that the project does not delay dues to lack of finance.

Cost analysis for each work is done to determine the quantity of raw material. Rate analysis captures the accurate quantity and cost of raw material, labour, transport, tools and plants, establishment and supervision and a reasonable profit. This also helps when there is a variation in the rates of materials. Projects of longer durations have an escalation clause. Escalation is the increase in the price of material originally quoted to the present cost. Quantity estimation is also necessary for material consumption and reconciliation especially of steel and finishing material.

The rules of calculating the quantities for various works are specified in IS1200. ISI published the standards of methods of measurements in 1958. Once the work is in progress, the measurements are captured in a notebook known as Measurement book or more popularly M-book. The system of units adopted across the world is the SI system. The rules of capturing the measurements and the deductions to applied have to be strictly followed as specified. One of the most common errors in quantity surveying is the unit in which the item is measured or expressed. For example brickwork is expressed in volume whereas the block work is calculated on surface area. Another common error done is quantity surveying is the deduction of openings. For structural works, opening or projections less than 0.5 m^2 are not considered for deductions. Some of the commonly used units are mentioned in Table 4.2:

Table 4.2: Table showing the units for various works

S.No.	Work	Unit	S.No.	Work	Unit
1	Excavation	cum	1	Formwork	sqm
2	Earthwork	cum	2	Blockwork	sqm
3	Concrete Works	cum	3	Half Brickwork	sqm
4	Full Brickwork	cum	4	Plastering	sqm
5	Reinforcement	Kg	5	Joineries	sqm
6	Skirting	rm	6	Flooring	sqm

DETAILED MEASUREMENTS						
		L	B	D	Qty.	Unit
1	BUILDING-CIVIL WORKS					
	EXCAVATION					
Footing	1	36	3.05	3.25	1.83	653.04 Cum
Footing	1	45	2.74	2.95	1.83	665.64 Cum
Raft	1	1	3.53	4.44	1.83	28.68 Cum
TOTAL EXCAVATION						**1347.35 Cum**
2	**RIVER SAND FILLING**					
Footing	1	36	3.05	3.25	1.83	653.04 Cum
Below Footing	1	36	3.05	3.25	0.15	53.53 Cum
Below Footing	1	45	2.74	2.95	0.15	54.56 Cum
Below Raft	1	1	3.53	4.44	0.15	2.35 Cum
Below Plinth Beam	1	7	22.86	0.20	0.15	4.80 Cum
Below Plinth Beam	1	2	13.56	0.20	0.15	0.81 Cum
Below Plinth Beam	1	5	29.26	0.20	0.15	4.39 Cum
Below Plinth Beam	1	8	45.52	0.40	0.15	21.85 Cum
Below Flooring in Basement	1	1	4.86	30.67	0.15	22.36 Cum
TOTAL SAND FILLING						**164.65 Cum**
3	**PIT SAND FILLING**					
Footing	1	36	3.05	3.25	1.83	653.04 Cum
Below Flooring in basement	1	1	4.86	30.67	0.60	89.43 Cum
TOTAL PIT SAND FILLING						**164.65 Cum**

Fig. 4.5: Figure showing a measurement sheet

6. CONSTRUCTION EQUIPMENT

The increase in sizes of construction has made it necessary to involve mechanical means to speed up the construction process. Some of the commonly used construction equipments used is as below:

6.1 *For Excavation*

(a) **Excavator:** Hydraulic excavators are heavy construction equipment consisting of a boom, a dipper, a bucket and cab on a rotating platform known as a house. The house sits on top of an undercarriage with tracks or wheels. Generically it is more commonly known as poclain, which is the company which popularized mechanical excavators.

(b) **Back Hoe Tractor:** Popularly known as backhoe, it consists of a tractor like unit fitted with loader style shovel or bucket in the front and a backhoe at the back. The seat can swivel to 180 degrees for the operator to face the side which is being used. Due to its popularity, this is also known as JCB in a generic term.

6.2 *For Transportation*

(a) **Wheel Barrow:** A wheel barrow is hand propelled equipment with just one wheel. It has two handles and is guided by a single person and it is used to transport heavier loads with ease. It is a second class level.

(b) **Builder's Hoist:** Builders hoist or a construction hoist is equipment which can be used to transport materials to higher floors. The construction hoist is made up of either one or two cars (cages) which travel vertically along stacked mast tower sections. The mast sections are attached to the structure or building every 7.62 m for added stability. For precisely controlled travel along the mast sections, modern construction hoists use a motorized rack-and-pinion system that climbs the mast sections at various speeds.

(c) **Concrete Mixing Transporting Truck:** Concrete mixing transporting truck or in-transit mixer is used to transport concrete of higher quantities from the batching plant to the site. The concrete mixing transport truck maintains the material's liquid state through agitation, or turning of the drum, until delivery. The interior of the drum on a concrete mixing truck is fitted with a spiral blade. The quantity of concrete which can be transported in one truck is 6 cum.

(d) **Tower Crane:** Tower cranes are a familiar sight in today's construction process. Tower cranes give a good mobility and access to location with difficult approach. They are also helpful in transporting heavy load in a short period of time.

6.3 For Concreting

(a) **Concrete Mixer:** Concrete mixer is a mechanical device use to produce concrete at site. It is used for smaller to mid range quantity of concreting. These mixers consist of a rotating drum mounted on an operator-driven cab-mounted chassis frame fitted with a loading bucket.

(b) **Needle Vibrator:** These are also known as immersion vibrators. It has a steel tube, called a poker, with one end being closed and rounded. There is an eccentric vibrating element inside it. The poker is connected to an electric motor, sometimes a diesel motor, through a flex tube.

(c) Needle vibrators come in a variety of sizes from 40 to 100 mm in diameter. The poker's diameter is determined by the spacing between the reinforcing bars in the formwork. The general range of vibrations for a needle vibrator is between 3000 to 6000 rpm. The period of vibration necessary can be from 30 seconds to 2 minutes and the concrete should be placed in layers no more than 600 mm high.

(d) **Batching Plant:** A concrete batching plant is used to produce concrete in large quantities. These plants are automated to assist is fast an accurate measurement of input of material. Because of the high degree of control in production, it is now feasible to produce high strength concrete.

Fig. 4.6(a): Concrete mixer

Fig. 4.6(c): Wheel barrow

Fig. 4.6(b): Builder's hoist

Fig. 4.6(d): Excavator

Fig. 4.6(e): Tower crane

Fig. 4.6: Figure showing construction equipments

Reference and Further Reads

Allen, Edward and Joseph Lano, "Fundamentals of Building Construction: Materials and Methods", 6th Edition, 2013, Wiley.

Chudley, R. and R. Greeno, "Building Construction Handbook", 7th Edition, 2008, BH, Elsevier.

Dutta, B.N., "Estimating and Costing in Civil Engineering—Theory and Practice", 28th revised edition, 2016, UBS Publishers' Distributors Pvt. Ltd. IS1200—parts 1 to 28.

SP 62 – Handbook on building construction practices

SP 20 – Handbook on masonry design and construction

SP 34 – Handbook on reinforcement detailing.

5

PROJECT MANAGEMENT

The word Project management consists of two parts, namely Project and management. A PROJECT is defined as a process, undertaken to create a unique product or service which brings about beneficial change or added value. A project has a specific start and completion dates. MANAGEMENT is the PROCESS of planning, organizing, and managing resources to bring about the successful completion of specific goals and objectives. Combining both the definitions we can say that Project Management is the process of planning, organizing, and managing resources to bring about the successful completion of specific project goals and objectives. The managing of various aspects of a project namely cost, time and personnel comes under this purview. The increasing sizes of projects and their complexities have increased the importance of this process and specialized teams are established.

In earlier days, design consultants and the contractors shared the responsibility of managing the project, the major onus of driving the project was with the contractor. In construction, the term construction management and project management overlapped and more or less meant the same. However, with stakeholders being more involved and the need to have control over the project, they either developed their internal teams to handle the project or hired project management consultants who would be the representative of the stakeholders and drive the project and coordinate all the teams and process involved in the project. With ISO standards being established for quality and safety, the implementation of documentation processes and the need to improve the processes became a primary concern in the project management. Thus the field of project management differentiated itself from the field of construction management.

1. HISTORY

The history of project management can be traced back to its roots in the planning process and attributed to Henry Gantt who is known as father of planning. His work on planning techniques and development of Gantt chart is basis of almost all project management tools used today. 1950's is considered the beginning of project management era in United States and projects were managed ad-hoc using mainly Gantt chart and some customized tools. Subsequently, two mathematical project scheduling models namely the "Program Evaluation and Review Technique" or PERT, developed by Booz-Allen & Hamilton as part of the United States Navy's (in conjunction with the Lockheed Corporation) Polaris missile submarine program; and the "Critical Path Method" (CPM) developed in a joint venture by both DuPont Corporation and Remington Rand Corporation for managing plant maintenance projects set the base for formalization of Project Management as a separate field.

2. FUNCTIONS OF A MANAGER—POSDCORB

POSDCORB is a widely used acronym in the field of management and public administration. This term gets prominence in a paper presented by Luther Gulick in 1937. The acronym stands for Planning, Organizing, Staffing, Directing, Coordinating, Reporting and Budgeting. Originally, Gulick proposed this in the explanation of his question on the role of a chief executive, but this could be applied to any manager.

(a) **Planning** is working out in broad outline of the things that need to be done and the methods for doing them to accomplish the purpose set for the project. This is the first and the most important activity of a management process. A good planning goes a long way in easing out the subsequent activities. This part defines the "what" of the project.

(b) **Organizing** is the process of establishing of the formal structure of authority through which work subdivisions are arranged, defined, and coordinated for the defined objective. This part defines the "how" of the project.

(c) **Staffing** is the whole personnel function of bringing in and training the staff and maintaining favorable conditions of work to ensure that the project has adequate manpower both in terms of numbers and qualifications. This part defines the "who" of the project.

(d) **Directing** is the continuous task of making decisions and embodying them in specific and general orders and instructions. This process is like navigation where the leaders of the project formulate the path to be followed. Proper direction can go a long way in successful completion of the project.

(e) **Coordinating** is the most important duty of inter-relating the various parts of the work. A project has various different personnel and activities and coordinating between them is of utmost importance to ensure a smooth progress and completion of the project.

(f) **Reporting** is the process of keeping all the relevant people involved in the project updated. The managers have the responsibility to keep the people informed as to what is going on to whom they are responsible. They also need to ensure that they and the people for whom they are responsible are also kept informed through records, research, and inspection.

(g) **Budgeting** does not just refer to the cash flow but encompasses all the planning of flow of material and labour. This is a critical area since this can have major implications in completion of a project. The variations in the costs of material and labour due to delay or price escalation can sometimes put the project under jeopardy. There can also be serious implications if the quantities of materials are incorrectly procured.

3. STAGES OF A CONSTRUCTION PROJECT

Once a project is initialized and decided to be implemented, there are various activities which are to be followed in a sequence. Figure 5.1 shows the process of how a project is implemented. Various activities in the entire process of a project from its conceptualization to delivery are Design, Planning, Implimentation, Delivery, and Post Project Evaluation.

3.1 Design

This is the first stage where the project is conceptualized and designed on paper. Inputs are taken from the stakeholders

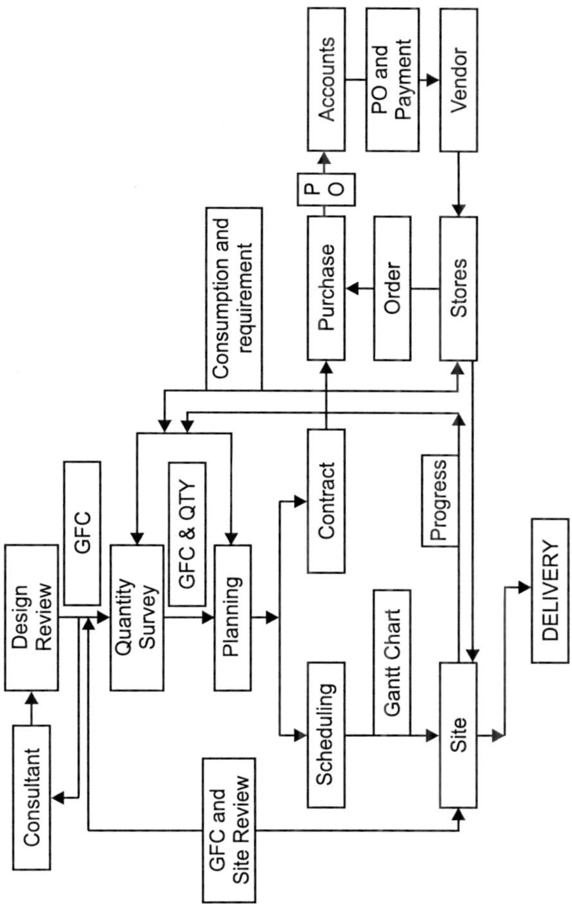

Fig. 5.1: Figure showing work flow in a construction project

and the research team and then the consultants namely the architects and structural engineers who are given the feedback of the requirements from the stakeholders. The designers come up with designs which undergo various iterations to satisfy all requirements of the project. Once the design is finalized, it is sent to the planning stage. The drawings are generally called as Tender drawing if it is for tendering only and Good for Construction (GFC) if it is ready for implementation. Normally GFC drawings are more detailed than the tender drawings.

3.2 Planning

The planning stage has three components—Quantity surveying, Scheduling and Contacting. This stage can also be called the project control as the control of the time and cost of the project is done here. This process is the crux of project management in construction industry and differentiated from construction management which deals in management of the project while being executed.

3.2.1 Quantity Surveying

Once the design is finalized the drawings are sent to the quantity surveying team which will estimate the quantity of work and the materials from the drawing specifications. This document is termed as Bill of Quantities or on short BOQ. This is the most important step for both budgeting and scheduling. Any errors at this stage would affect the project in totality and hence the accuracy at this stage should be very high. From here the drawings and the BOQ are sent to the tendering or contracting department for preparation of tender papers for selection of contractors and to the scheduling department to ascertain the time for the project. Fig. 5.2 shows a sample BOQ.

3.2.2 Scheduling

Scheduling is important for budgeting and cash flow. Proper scheduling cannot only determine the time lines for project completion but also help to ensure that the project does not delay due to the lack of cash flow. The most common tool for scheduling is the Gantt chart or bar chart. Bar chart shows the start and end of each activity and also shows the overlaps of start of each activity. It is very important to know and understand the sequence of activities and the labour turnover for each activity. Fig. 5.3 shows a bar chart for construction project.

3.2.3 Contracting

Preparation and awarding of contracts is the next stage after quantity surveying and scheduling. The contract gives a legal binding agreement between the stakeholder and the contractor. Every country or region had their own specification of contract and in case of dispute, the arbitration was under the jurisdiction of the region where there the contract was being executed. When the industry started turning global and players from different region or countries started taking up contracts in countries other than their own it became a tedious process to ensure awareness of local laws. This paved way for a more standardized conditions of contract applicable uniformly. The international federation for consulting engineers (FIDIC) established a uniform contracting template for different kind of works.

S. No.	Description	Qty.	Unit	Rate in ₹	Total in ₹
I	**BUILDING–CIVIL WORKS**				
"A"	**EARTHWORK**				
1	Earthwork excavation in all types soil including sub-soil back filling the foundation trenches with excavated earth in layers not exceeding 150 mm in depth, watering, consolidation and disposal of surplus earth as directed below existing ground level etc. complete.	47.98	Cum	0.00	0.00
2	Supplying and filling in foundation and basement with river sand depth of layers not exceeding 6″ in depth, including watering, consolidation etc., complete	29.98	Cum	0.00	0.00
3	Supplying and filling in foundation and basement with pit sand depth of layers not exceeding 6″ in depth, including watering, consolidation etc., complete.	147.46	Cum	0.00	0.00
	Total "A" Excavation carried over to summary				**0.00**
"B"	**CONCRETE**				
1	**PLAIN CEMENT CONCRETE WORKS**				
a	Plain cement concrete (1:4:8) using 40 mm and below size broken stone jelly in leveling course under floor including machine mixing, formwork if necessary, ramming, consolidation, curing etc, complete below footings, plinth and capping beams and in basement below flooring	34.68	Cum	0.00	0.00

Fig. 5.2: Figure showing sample BOQ

S.No.	Work Description	Week 1	2	3	4	5	6	7	8	9	10	11	12	13
1	Excavation	■												
2	Sand filling below foundation	■												
3	PCC below foundation	■												
4	Laying of reinforcement for footing		■											
5	Placing of side forms			■										
6	Concreting of footings			■										
7	Casting of starter blocks			■										
8	Raising of columns to plinth level				■									
9	Back filling of foundation				■									
10	Sand filling below grade beam					■								
11	PCC below grade beam					■								
12	Casting of grade beam						■							
13	Raising of blockwork to plinth level							■	■					
14	Filling of plinth/basement									■	■			
15	Laying of PCC at plinth level											■		
16	Raising of column to roof level												■	

Fig. 5.3: Figure showing a sample bar chart for scheduling

(a) **Types of Contract:** There are three main types of contract lump sum contract, item wise or schedule contract and cost plus contract. There are other types of contracts which are a variation or combination of these three types.

(i) **Lump sum Contract:** In this type of contract, a total fixed price is arrived at for the total contract for the required specification. This is the most common type of contract used. This type of contract is preferred when the scope of work is clear and defined. Another reason for the stakeholders to prefer this type of contract is the easy understanding and when the risk of changes and additional unspecified works needs to be transferred to the builder. However the contractor on his part considers these factors and quotes his price. The drawback of this contract is in case of dispute or contractor leaving midway, it is very difficult to determine the value of work.

(ii) **Item/Unit-wise or Schedule Contract:** This is an ideal contract where there is a transparency on both sides. In this type of contract, the unit price and specification of each activity of work is mentioned and contractor quotes for the quantity calculated. The advantage of this type of contract is that the contractor is paid only for the quantity of work completed. In case of dispute, there is clarity in the work completed and payments to be made. All major works today follow this type of contract.

(iii) **Cost Plus Contract:** In this type of contract, the actual cost of the work is calculated and contractor is paid a fixed percentage for all the works over and above the cost for managing the project. This type of contract is generally carried out when the client supplies all the resources and the contractor is more of a manager. This type of contract is appropriate when the risk on delay of project is accepted by the client. This type of contract is also executed when the data available is insufficient to estimate the final cost of the project.

(b) **Contract Document:** Contract document is a legal binding document drawn between client and the contractor. There are three parts to a contract document—general conditions, technical specifications and bill of quantities.

  (i) **General Condition:** This part consists of mainly the articles of agreement. This part describes the definitions used, description of project, type of contract, assignments of roles and responsibility, working conditions and jurisdiction in case of disputes.

  (ii) **Technical Specification:** This part describes the technical specification and descriptions of all the material to be used and the process to be followed during the execution of the project. The technical specifications are generally adopted from IS standards in India.

  (iii) **Bill of Quantities:** This part gives the quantities and specification of the items of work to be executed in the project and the price quoted by the contractor.

(c) **Tendering:** Tendering is the process of bidding and awarding a contract. Depending on the size and importance of the project, the tendering process can be done in one stage or two stages. In the two stage process, stage 1 is a process called as pre-qualification. In this process, the selection is done based on the capability both technical and financial and market standing of the organization to carry out the contract based on some past details. This stage is also sometimes known as pre-bid. The details are generally asked in form of a questionnaire, viz:

- Company details including the legal status of the company.
- Financial information such as recent accounts and the insurance covers.
- Relevant experience with details of past projects executed.
- Technical and professional capability.
- Management policies including quality and health and safety.
- References from past projects.

The second stage is the actual bidding where the selected contractors from stage 1 quote the financials to execute the project. All things being equal, the tender is generally awarded to the lowest quoted price unless there is an anomaly in the pricing. However in certain cases lowest bidder is not awarded the contract. The financials are also evaluated for viability. Even if the contractor has quoted the lowest, if the evaluators feel that the quote is impractical, the contract is awarded to the next best option.

3.3 Implementation

Once a contractor is finalized, the process of implementation begins. From the contractor's side, this part is generally defined as the construction management as this manages the actual construction. During the implementation stage there will be iterations on planning where there are possibilities of changes is design, costs and time due to various circumstances. Though this variance is not desired, however there is rarely a project which has not undergone changes and has been implemented as per plan.

3.3.1 Construction Management

The contractor assigned to a construction project has to diligently ensure that the project is implemented as per the requirements. The construction management team, or the site team as it may be termed, implements the project and monitors the progress on site. Constant feedback has to be taken from the site team on the current position of the project. Any hindrance encountered should be intimated back to the design team and the project control team. The feedback from the site on the actual prevailing condition is necessary to ensure that the project runs smoothly.

The site team and the construction manager have to be well versed in the construction process and sequence. Starting from the site preparation to finishes (partly explained in chapter 4), the proper sequence should be followed. Apart from the sequence and progress, the site team needs to look into ways and means by which the construction can be carried out within stipulated time and cost for maximum benefit without compromising on the ethos of the project.

3.4 Delivery

Delivery is the final phase of the construction where the project is ready to be handed over to the client for intended use. This part is also called the handover. Although the project close-out and handover are typically the final phase of the project this does not mean that the relevant activities should only commence when the previous stage is complete. On the contrary, it can be seen by the list above that work such as built drawings should be developed as the project progresses through the earlier stages and be ready for handover as soon as the work is complete.

The delivery can be a single stage activity or a two-stage activity depending on the end user. If the stakeholder is the end user then the delivery is a single stage process. However in many projects, especially in development projects, where the client and end user are different it is a two-stage process. The completed project is delivered by the contractor to the developer and the subsequently the developer hands over the project to the end user. A typical example is the large residential projects developed by the promoters.

Once the project delivery is completed, the financial closure needs to take place. Financial closure is the stage where it is seen that all the conditions related to financing agreements are fulfilled.

Finally, on the completion of the project, it is to be ensured that the project team members are reassigned; surplus equipment, materials and supplies disposed off.

3.5 Post Project Evaluation

The final step in any project should be an evaluation review. This is a look back over the project to see what was learned that will contribute to future projects. This review is best done by the core project team and typically as a group discussion. If possible it can be beneficial to include the client, customer, users or other stakeholders. The post-project evaluation review will take place during project closure.

3.5.1 Need for Evaluation

Rarely a project has been completed as planned. With all projects there are areas that can be improved upon; however, it is typical for project personnel and managers, to get away from the project as soon as possible. This may be due to the fact that they are required immediately on the next project and they are glad to see the back of it and want to forget the results. Whatever be the reason, the benefits of such a review are rarely seen, and dismissed as time-wasting. It is therefore very important that the lessons learned from the project, whether good or bad should be reviewed, documented and fed back to the next project. Without a proper evaluation, work will be carried out in the same way and the same mistakes will be repeated in every project.

3.5.2 Evaluation Topics

It is very important to conduct an evaluation of a project delivery against the expectations. This highlights the importance of agreeing the measurement of success at the outset of the project and with all stakeholders. Before the start of the project, there are three critical questions which need to have a positive answer. They are:

- Do we know where we are going?
- Do we know how we are going to get there?
- Do we have agreement of all of the stakeholders on the first two questions?

On completion of the project the same three questions need to be evaluated:

- Did we get where we expected to get?
- Did we get there the way we expected to?
- Are all the stakeholders happy with the outcome and the way it was achieved?

The type of answers to the above question decides the depth of evaluation which needs to be done for further improvement.

3.5.3 Correction Measures

If the answer is no to any of the above questions we should be asking why and what can be done to make sure the answer is yes in the future. Equally, if the answer is yes we should be congratulating the project team and ensuring that we clarify what was done well so we can do it again and also what could be improved upon next time.

Some questions that could be considered are:

- Was the project direction and objectives understood and agreed universally and carried out as planned?
- Was there an optimum and appropriate use of skills, tools and techniques in the project?
- Was there cohesiveness in the team with regards to sharing of responsibility and openness in feedback?
- Was there a transparency on the project with the stakeholders?
- Was project challenges anticipated and managed effectively?

Again if the answer to any of these questions is no, then there should be an exploration on the reasons for the gaps. Appropriate measures should be taken to ensure that in the future they are corrected. Likewise, if the answers are yes then the reason for the same should be explored. It should be ascertain if it was due to the correct method followed or there were external unexplained events. Either way, the results should also be documented and fed back into future projects.

4. VALUE ENGINEERING

Value engineering is the process of evaluating a process or product and suggesting methods to replace the material or improve the output without compromising the function, quality and budget. Lawrence Miles, an employee of GE during the Second World War is considered the pioneer in suggesting a method for value engineering. In value engineering, value is defined as function divided by cost. Value can be increased by either improving the function or reducing the cost. Concentrating on the maximizing the project's function while minimizing cost generates the greatest value for clients.

Application of value engineering in a construction project has enormous benefit for developers. The multi-step process needs to be an integral part the design stage of a new development. It is used to identify and eliminate unwanted costs. The process involves looking into all the factors which have direct or indirect effect on the project. Consideration should be given to factors like availability of material, construction methods, planning and organization etc.

Value engineering should ideally start at the inception stage. However during the course of the project the contractor can make significant contribution if it does not affect the timelines or costs. From the contractor's side, the project manager should play a pro-active role in leading and directing the value engineering process.

4.1 Steps in Value Engineering

There are four basic steps in value engineering:

(a) **Information Gathering:** This step evaluates the requirements of the object or process. Function analysis, an important technique in value engineering, is usually done in this initial stage. It determines what functions or performance characteristics are important and what can be compromised. The answers to the following questions need to be looked into,

- What can or should the object do?
- What must the object do or what it must not?

(b) **Alternative Generation and Creation:** This step looks for alternatives which will perform the same function of the object for which value engineering is carried out. The value engineers have to find answers to the two questions,

- What are the various alternative ways of meeting requirements?
- What else will perform the desired function?

(c) **Evaluation:** In this stage all the alternatives are assessed by evaluating how well they meet the required functions and how great will the cost savings be.

(d) **Presentation:** In the final stage, the best alternative will be chosen and presented to the client for final decision. The presentation should clearly state and discuss why the particular option was chosen or opted for ahead of all the other feasible options and how it affects the project and the stakeholders.

5. RE-ENGINEERING

Re-engineering is defined as the fundamental rethink and radical redesign of processes to generate dramatic improvement in critical performance measures such as cost, quality, speed and service. The concept of re-engineering in construction started in late 1990's. The constant evolution of construction industry has necessitated the engineers to look at innovations. Each project presents a unique set of problems which need to have specialized approach. The processes need to have a relook and redesigned. The traditional ways and approaches have been questioned and relooked into.

Betts and Wood Harper in 1994 stated that the construction industry looks to comfortably adopt local practices rather than seek to find the best global practices. Despite its advances in field of civil engineering around the world, majority of India still follows the traditional processes and techniques. There is reluctance to adopt new materials and processes. People find themselves stuck in deep-rooted cultural habits and behavior and continue to follow them without understanding the underlying reasons. Even though it is a challenge to get people to embrace new ways of thinking by understanding the existing systems, time is ripe to introduce and accept changes to be on par with the changing world scenario.

The concept of re-engineering in construction industry is still under evolution and relatively new and it needs to be adopted in a much larger way to integrate and implement the concept in the construction process.

Reference and Further Reads

Dell'Isola, Alphonso, "Value Engineering: Practical Applications...for Design, Construction, Maintenance and Operations", 1997, Rsmeans.

Ganaway , Nick B., "Construction Business Management", 2106, Routledge.

Hammer, Michael and James Champy, "Reengineering the Corporation: A manifesto for Business revolution", 2006, Harper business.

Jackson, Barbara J., "Construction Management Jumpstart", 2nd Edition, 2010, Sybex.

Kumar Neeraj Jha, Construction Project Management.

Mukhopadhyaya, A.K. *Value Engineering: Concepts, Techniques and Applications.*

Soares, Roberto, "Reengineering Management of Construction Process", 2013, *International Journal of Business and Social Science,* Vol. 4, No. 7.

6 QUALITY

Quality is defined as the process where the product or the process is maintained as per the prescribed standard which will give the optimal output of the desired result. According to ISO8042—1986, quality is defined as "the totality of features and characteristics of a product or service that bears its ability to satisfy stated or implied need". Generally quality is the agreed standard between all the people involved with the product or process, be it producer or the end user.

In the earlier days the onus on maintaining quality was inbuilt and the craftsmen ensured that the product delivered was to the desired requisition and any deviation had a reasonable explanation. However with the increase in product and people and processes getting more complicated, quality became a field by itself. People were hired to specifically look into quality. Eventually it became a department and field by itself.

In construction, quality plays a very important role. Adherence to quality has both tangible and intangible benefits. The tangible benefits are the savings in cost and time involved in maintaining quality and further repairs in case of non-compliance. The defects and their repairs can result in very large cost which can affect the profit from the project and the time taken for rectification can delay the project. With time being a criterion in the today's fast pace world, penalties are imposed on the contractor for delays in work. The intangible benefit is the brand image which will further affect the standing in the market and future expansion of the organization or business at large. Hence it is beneficial in more than one ways to ensure that the quality is maintained at the first instance and ensure that there are no reworks.

1. QUALITY STANDARDS

The turn of the 20th century opened a new chapter in material specifications. Industrial revolution brought about the invention of new materials and techniques. The invention of Bessemer's process made steel the most widely used material. Large quantities of steel were required for setting up railroads across the world. Simultaneously there was an increase in use of concrete in construction. The absence of standard specifications and high failure rate of materials made it difficult for buyers and end users of materials to decide on the appropriate material. This dilemma gave birth to the idea of creating a uniform specification and establishment of a governing body to decide, enforce and monitor standards. Originally the governing bodies were set up for testing and certifying materials. Subsequently the processes were also brought under the purview of the quality standards. As is the case with any new system, there was resistance in acceptance of new standards and procedures. With proper education on the larger and long-term benefits, a consensus was achieved in implementation of the standards.

1.1 ASTM

Initiated by Pennsylvania Rail road, Charles Benjamin Dudley was the driving force in finding a solution to the steel and other material used in railroad. In 1878, he published his first major report, "The Chemical Composition and Physical Properties of Steel Rails," which described the durability of different types of steel rails. In 1898, Dudley and a group of scientists and engineers laid foundation to an organization which was called ASTM (American Society for Testing Materials). In 1961, the name was changed to American Society for Testing and Materials. In 2001, the name was changed the presently used ASTM International.

The membership to ASTM is open to anybody and as on date there are over 30000 members from over 140 countries. ASTM's world headquarters are located in West Conshohocken, Pennsylvania, with offices in Belgium, Canada, China, Peru and Washington, D.C. There are over 13000 standards published. In 1902, the Committee C-1 was founded to standardize test methods in the cement and concrete sector.

1.2 British Standard

Like ASTM, British standard institute was formed in 1901 to standardize the manufacturing of steel sections. Led by Sir John Wolfe Barry, the first standard was published by BSI was for steel sections used for tramways. BSI created one of the world's first quality marks in 1903, when the letters 'B' and 'S' (for British Standard) were combined with V (for verification) to give birth to Kitemark logo. BSI pioneered the development of management systems standards. BS5750, the original quality management systems standard formed the template for the ISO9000 series of quality standards and was first published in 1994.

At present BSI have 27000 active codes. BS8110 was the british standard developed for design and construction of reinforced and pre-stressed concrete. In 2010 BS8110 was superseded by EN1992 (Euro code 2).

1.3 Indian Standard

Indian Standard Institution was established in 1946 by the Department of Industries and supplies to take up the task of formulation of national standards. The Institution of Engineers (India) formulated the first draft of the constitution. ISI formally came into being on 6th January 1947 with Dr. L.C. Verman as its first director. Originally set up for establishing standards, ISI introduced the scheme of issuing certification marks for the products confirming to the standards established by ISI. Certified testing laboratory was started in 1963.

As was the case with other organizations, at first the role of ISI was limited to product standardization only. In 1986, ISI was upgraded to BIS with a broadened scope which included processes too. As on date there are over 19000 standards published by BIS. For civil engineering there is close to 1700 Indian standard (IS) codes and 20 Standard Practices (SP) codes. IS456—2000 is code of practice for plain and reinforced concrete.

1.4 ISO

With a view to have uniform global standards, 25 countries met in 1946 at the premises of The Institution of Civil Engineers, London to set up a global standard body. The role of the governing body was to facilitate the unification of industrial standards across the world. On 23rd February 1947, ISO (International Standard Organization) formally began its operations with its headquarters in Geneva, Switzerland. As on date ISO has published over 21000 international standards covering almost all aspects of technology and manufacturing.

In 1987, ISO published ISO9000, which became the foundation for Quality Management principles. It is based on the seven quality principles namely, Customer focus, leadership, Engagement of people, Process approach, Improvement, Evidence based action and relationship management. Though ISO itself does not certify, there are certification bodies who audit organizations for ISO9000 compliance. The organizations are then issued ISO9001 compliance certificates.

2. QUALITY COMPLIANCE IN CONSTRUCTION

Compliance is the adherence to quality and in construction it ranges from the standards involved in the procurement of raw materials to the process and the finally the finish. Indian Standard code clearly defines the standards to be maintained in terms of material and process. It is imperative in today's work condition that the contractors submit their quality plan and same is vetted by the specialized quality team consisting of engineers and consultants. In the quality plan, for materials, the quality plan should consist of the types of test to be conducted on the materials and their frequency as per specifications. The quality plan for procedures should consists of the documents to be maintained for each process carried out, like pour cards and further repair and rehabilitation techniques in case of variance from the standards on strength and finish. Fig. 6.1 shows the sample page from a QA plan indicating the frequency of testing. Fig. 6.2 shows a sample pour card used during the construction stage.

Contractor:				QA/QC MATRIX	
Project:					
Client:					
Consultant:					
S.No.	**Description of Material/Item**	**Tests to be conducted**	**Accetable test results**	**Time and Interval**	**Reference code**
1	Ordinary Portland Cement	Initial Setting time	Not less than 30 mins		
		Final Setting time	Not more than 600 mins		
		Fineness	Not more than 10% by weight passing through 90 micron	5000 bags or change of brand	IS8112
		Compressive strength @28 days	As per grade of cement		
		Chemical Composition	As per manufacturers test report		
2	Reinforce-ment steel Fe415	% Elongation	Min 14.5%	Each Lot	ISI 786
		Proof strength	415 MPa		
		Ultimate Strength	485 MPa		
		Unit weight	As per ISI 786 Table 3		
		Chemical composition	As per manufacturers test report		
3	Fine Aggregates	Seive Analysis	Should confirm to zone 1 or 2	Every 100 cum or change of source	IS383
		Silt content	Max 8% by volume		
		Sieve Analysis	As specified in IS2386	Every 100 cum or change of source	IS2386
		Specific Gravity			
		Bulk Density			
		Crushing Strength			
		Abrasion resistance			
		Soundness			
4	Concrete Blocks	Density	Should not be less than 1800 kg/cum	Every 1000 blocks or change of source	IS2185
		Compressive strength	3.5 MPa for solid blocks		
		Water Absorbtion	Should not exceed 10%		

Fig. 6.1: Figure showing sample Quality Assurance Plan

INSPECTION SHEET (STRUCTURAL CONCRETE)			
CHECK LIST NO:		Sheet No:01	
Project			Date:
Location/Grid:		Column ID:	
Drawing Reference:		Grade of Concrete:	
Bottom Level:		Top Level:	
S.No.	**Description**	**Status**	**Remarks**
1	Layout as per Drawing		
2	Ref points/base line using total station		
3	Shuttering supports per approved design		
4	Vertical of Props		
5	Screw/cup couplers		
6	Bottom and top bearing plates		
7	Height between plates, to restrict unscrewed length of jacks		
8	Arrangement of shutter plates/board		
9	Verticality of beam/col/CW sides		
10	Levels of top of shutter		
11	Condition of shuttering material		
12	Proper size of shutter at junctions of B-C, C-S, B-S		
13	Cleaning and Oiling for shuttering		
14	Sealing of shuttering joints		
15	Top of Slab marks at all cols/edges		
16	Bar bending schedule as per drawing		
17	Straightness of all reinforcement		
18	Column/Wall/beam/slab reinforcement in position		
19	Bearing of reinforcements at col/beam jn		
20	Development length of bars (lap. L.)		
21	Proper Cover, with blocks/runners/chairs.		
22	Dowels/inserts/sleeves/conduits e.g. Stair/ramp/beam		
23	Back Propping of 3 lowers floors 50%, 25%, 10%		
24	Approved construction joint in position		
25	Shear key/dowels		
26	Vertical stoppers at beams/CWs/slab		

Fig. 6.2: Figure showing sample Pour card/Check list

2.1 Testing of Materials

Testing of materials to be used is an important aspect of quality in construction. A good quality raw material will go a long way in ensuring the quality of final product. There are two types of tests conducted on materials namely Chemical test and Physical tests. The chemical tests determine the inherent composition of the material and the physical tests determine the behavior of material under various situations. The chemical tests are done in laboratory and the predominant role is played by a material sciences expert. The physical tests can be carried out in laboratory as well as on site.

RCC (Reinforced Cement Concrete) being the primary construction material, it is imperative that we lay stress on the quality of raw material used in production of RCC. The physical tests required for various materials used in RCC are as follows:

2.1.1 Testing of Cement

(a) **Field Testing:** For a quick result, to find out the quality of cement, field test can be conducted. This is a more of an approximate result and can be used for minor works.

Following are the steps in field test:

- The cement is to be visually inspected for any lumps and colour. The colour of the cement should be greenish grey.
- Once the hand is put inside the cement, it should give a cool feeling.
- When a pinch of cement is taken between the fingers, it should give a smooth feeling and not a gritty feeling.
- Stiff cement paste is to be prepared taking around 100 gms of cement and required quantity of water.
- The stiff paste is to be made into a cake with sharp edges by patting.
- The cake is to be placed on a glass plate and slowly immersed in water. The cake should not be disturbed while being immersed.
- The cake should retain its shape after 24 hrs of immersion.

(b) **Test for Determination of Compressive Strength:** This is the basic test which is carried out on cement. The compressive strength of concrete does depend on the compressive strength of cement too. The procedure is described in IS4031 part 6. The test specimen for this is a cube made of cement mortar mixed in a ratio of 1:3. The size of the mould is specified in IS10080.

Following are the steps involved:

- Standard sand, as specified in IS650 is used for preparing the cement mortar. The quantities of cement and sand are 200 gm and 600 gm respectively. The water to be added is based in the formula [(P/4) + 3] % of the total weight.
- Cement and sand are dry mixed to get a uniform mix and then the water is mixed and cement paste of uniform colour is made.
- Standard moulds of size of 50 cm^2 of surface area are prepared. The inner walls of the mould are coated with oil and the cement mortar is poured and prodded for about 20 times in about 8 sec. The mix is compacted by placing the specimen on table vibrator and vibrated for about 2 mins.
- The mould is then kept for 24 hrs in room temperature and at a relative humidity of about 90%. After 24 hrs, cubes are removed from the mould and submerged in freshwater for a period of 28 days. The water needs to be changed every 7 days.
- Minimum number of sample to be tested is 3. The testing of cubes is done by loading the cubes on the sides without any packing between the cube and the steel plates. The load is gradually increased to a maximum rate of 35N/mm^2/min.

(c) **Fineness Test:** Fineness has an important role to play in the evolution of heat of hydration and subsequently on the gain of early strength. Finer cements have higher rate of heat of hydration and hence gain strength earlier. There are mainly two methods to find out fineness namely:

(i) **Sieve Method:** Following are the steps to be carried out in this method:

- 100 gms of cement is taken and on a standard IS sieve no 9. IS sieve 9 means, the sieve is 90 μ in size.
- The sample is sieved continuously by giving it a circular and vertical motion for 15 mins.
- The weight of the residue on the sieve is found out and the percentage retained is calculated.
- The percentage retained should not be greater than 10%.

(ii) **Air Permeability Method:** Figure shows the apparatus used in this method. The apparatus contains following parts:

- Permeability cell (it is a metal cylinder in two halves).
- Manometer having kerosene oil.
- Flow meter filled with kerosene.
- Rubber tubes, capillary tubes.
- Perforated plates (sieve of dia 1.59 mm).
- Filter paper placed on perforated plate.

Following are the steps carried out in the test:

- The two halves of the permeability cell are bolted firmly and air tightness is ensured.
- Density of the cement is found out.
- A layer of cement of 1 cm thick is placed over the filter paper inside the cell.
- The cell is tapped 4 times by allowing the cell to fall for a height of 2 cms on a wooden bench.
- A plunger of required length is inserted to touch the top of the cylinder and moved slowly.
- Manometer is connected to both ends and rate of flow of air is adjusted in such a way the flow meter shows of difference in a range of 30 to 50 cms.
- The process is repeated till there is no variance in the ratio of h_1 and h_2.

- The specific surface is found out from the formula specified in IS269.

  $S = \lambda \sqrt{(h_1/h_2)}$ where $\lambda = \{14/[d(1-n \text{ or } \lambda)]\} * \{\sqrt{[(A\lambda^2)/KL]}\}$

  Where K = Flowmeter constant

  L = Thickness of cement layer

  A = Area of cement layer

  d = Density of cement

  n or ψ = porosity of cement.

(d) **Test to Find out Consistency of Cement Paste:** It is a test conducted to estimate the optimum quantity of water required to be mixed with cement to for the cement paste. The apparatus used to for the test is called Vicat's apparatus. Vicat's apparatus is shown in figure 6.3.

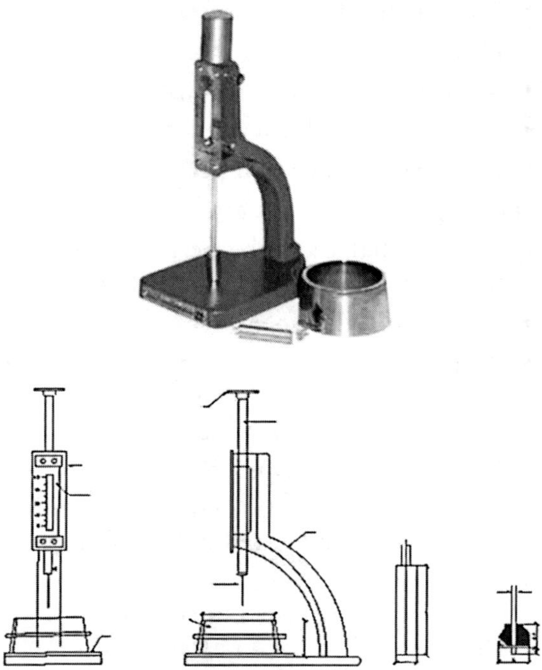

Fig. 6.3: Figure showing Vicat's apparatus

Vicat's apparatus consists of the following parts:

- Rod weighing 300 gms.
- Needle of length of 5 cm and cross-sectional area of 1 mm^2.
- Stand.
- Vicat's mould having a diameter of 80 mm and height 40 mm.
- Plunger of length 50 mm length and 10 mm diameter which is removable.
- A movable indicator attached to rod B.
- Non-porous plate to support the mould.

Following are the steps carried out in the test:

- About 300 gms of cement is taken by weight and mixed with 25% of water. The timing of mixing should be between 3 to 5 minutes.
- The cement paste is then filled in the mould of Vicat's apparatus. The time taken from adding the water to the cement and placing the cement paste in the mould is called the time of gauging and should be 4 minutes.
- The mould is placed under the rod B to which the plunger is attached. The plunger is lowered gently to touch the surface of test block and then quickly released and allowed to sick freely in the test specimen.
- The settlement is observed. If the settlement is between 5 – 7 mm from the bottom of the mould then the amount of water is adequate. If the settlement is outside this specified measurement, then the quantity of water of adjusted and test is repeated again.
- The test should be performed at room temperature only.

The percentage of water is calculated as ratio of weight of water required for the value of plunging required to the weight of cement.

(e) **Test to Determine the Soundness of Cement:** Soundness is the property of cement where it ensures that the cement is able to maintain constant volume, i.e. there is no appreciable change in volume of cement after the cement is set. The procedure to carry out this test is known as Le Chaterlier's method. The apparatus used is called Le Chatelier's apparatus and the procedure is described in IS5514. It is shown in figure 6.4.

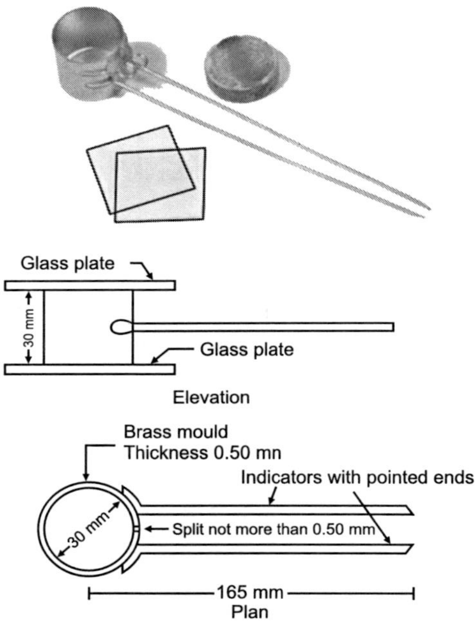

Fig. 6.4: Figure showing Le Chatelier's apparatus

The Le Chatelier's apparatus consists of the following parts:

- Split cylinder of spring brass: This is a cylinder which has a split of 0.5 mm as shown in figure. The cylinder is 0.5 mm thick and 30 mm high and has an inter diameter of 30 mm.
- Indicators: There are two indicators of 2 mm in diameter made of brass wires and braced on the either side of the cylinder. The distance beween the free end of the wire and centre of cylinder is 165 mm.

Following are the steps carried out in the test:

- 100 gms of cement is taken and mixed with water to form a paste. The quantity of water should be 78% of the weight of water required for standard paste.
- The mould is place on a glass plate and cement paste is filled in the mould and then covered on the top with another glass plate. A small weight is placed over the top glass to ensure that the cover does not displace.
- The assembly is placed in a water bath at room temperature and kept for 24 hrs.
- The assembly is removed from water bath after 24 hrs. and the distance separating the indicators is measured.
- The mould is then immersed in water and the water is heated to boiling point. The boiling of water is done for 3 hours.
- The mould is removed from water and then cooled and then the distance between the indicators is measured.
- The distance between the two measurements is the value determining the soundness of cement. The value should not exceed 10 mm for a cement to have good soundness.

(f) **Test to Determine the Tensile Strength:** Tensile strength is the ability of the material to withstand tensile load. The tensile strength of cement is determined on a specimen of cement mortar and subjecting it to direct tension. The mould is called briquette and the testing machine is called briquette testing machine.

Following are the steps to carry out the test:

- Dry Cement mortar is prepared with cement weighing 250 gms and standard sand weighing 750 gms.
- Water of [(P/2) + 2.5]% of total weight of dry mortar is taken where P is the percentage of water required for standard consistency.
- Wet cement mortar of uniform colour is prepared using the dry mortar and water.

- Mortar is place in the briquette and moulds are prepared. 12 moulds are prepared out of which 6 are tested after 3 days of curing and the remaining 6 are tested after 7 days of curing.
- Specimen is tested as shown in figure. The loading is done by allowing the lead shots to fall from hopper to the pan. Due to the lever action, the mould is imparted with tensile force.
- The supply of the lead shots gets automatically cut off when the mould fractures. The lead shots in the pan are weighed. The ratio of load to the cross-sectional area is the tensile strength of the cement.
- For OPC, the value of tensile strength should be 2 MPa after 3 days and 2.5 MPa after 7 days. For rapid hardening cement, the value after 7 days should be 3 MPa.

2.1.2 Testing of Aggregates

(a) **Test to Find out Fineness Modulus or Sieve Analysis:** Gradation is defined as the distribution of various sizes of particles in a bulk amount of aggregate. This is determined by sieve analysis. This test is specified is IS2386 part 1.

Following are the steps to carry out the test:

- The weight of the sample to be taken is specified in Table IV.
- The sample should be air dried and then the actual weight of the sample is determined.
- The sample sieved through various sizes of sieves. The standard sizes of sieves are 80 mm, 40 mm, 20 mm, 10 mm, 4.5 mm, 2.36 mm, 1.18 mm, 600 μ, 300 μ, 150 μ.
- The weight of the residue on each sieve is determined and percentage of retention is determined. The result is reported as the cumulative percentage by weight of the total sample passing through each sieve.

(b) **Test to Find out the Bulk Unit Weight:** Bulk unit weight is the total weight of aggregates. This is a very important value in mix design of concrete.

Following are the steps to carry out the test:

A cylinder of size 15 cm in diameter and 17 cm in height is thoroughly cleaned and dried and its empty weight is found out (W_1).

It is filled with and in 3 layers and each layer is compacted with a tamping rod of 16 mm diameter and 60 cm in length.

The filled cylinder is weighed (W_2).

The unit weight of sand is found out by the formula

Υ = (Weight/Volume) = $(W_2 - W_1)/(\pi r^2 h)$

The experiment is repeated and the average is taken as the unit weight of the lot.

(c) **Test to Find out Specific Gravity of Coarse Aggregate:** Specific gravity is also a very important value in mix design of concrete. The procedure for the test is specified in IS2386.

Following are the steps to carry out the test:

- Sample of Coarse aggregates is taken. The quantity of sample is 2 kg by weight.
- It is thoroughly cleaned by placing in a wired basket and immersed in distilled water in room temperature.
- The entrapped air is removed by constantly dropping and raising the basket from the water for 25 times.
- The sample is kept immersed in water for 24 hrs and then the sample is weighed (W_1).
- After removing the sample from water, the basket is emptied, dried and weight of empty basked is found (W_2).
- The aggregate is weighed separately in air (W_3).
- The aggregate is heated in oven for 24 hrs at a temperature of 100 to 110.
- The weight of the dried aggregate is found out (W_4).
- The specific gravity and water content are found out by the formulae given below:

  Specific Gravity = $W_4 / [(W_1 - W_2) - (W_3)]$

  Water content = $(W_3 - W_4)/ W_4$.

(d) **Test to Find out the Crushing Strength:** Crushing strength of aggregate ensures the compressive strength of concrete. The procedure for the test is specified in IS.

The steps to carry out the test are as follows:

- The total quantity of aggregate is weighed.
- The aggregate is place in a cylindrical mould.
- The material to be tested is divided into three lots.
- Each lot is tampered 25 times with the help of tampering rod.
- The material crushed to finer particles (less than 2.36 mm size) is sieved and separated and is weighed. The weight is expressed in the percentage of the original weight of the aggregates tested.
- The percentage is known as the aggregate crushing value.
- The crushing value of aggregates should be less than 30% for pavement works and 45% for other concrete works.

(e) **Abrasion Resistance Test:** The concrete used for pavements will undergo abrasion or wear and tear due to the movement of traffic. The aggregates used should ensure that the concrete has requisite strength to withstand abrasion. There are mainly three types of test to determine abrasion.

(i) **Deval's Attrition Test:** In this test, the sample of aggregates to be tested are subjected to wear and tear by placing the sample in a iron cylinder and rotated 10000 times.

(ii) **Dorry Abrasion Test:** This test is not mentioned in Indian standard. In this test the sample is placed in a cylinder of size 25 cm in diameter and 25 cm in height. The specimen is subjected to abrasion by rotating a metal disc in the cylinder for 1000 revolutions. The hardness is determined by the formula

$$[20 - (\text{loss in gm}/3)]$$

If the value of hardness is less than 14, then it is poor.

If the value of hardness is more than 17, then it is good.

(iii) **Los Angeles Test:** This is the most appropriate test for determining the abrasive strength of aggregate.

The steps followed in this test are as below:

- The aggregates to be tested are sampled as specified in IS2386.
- The Los Angeles testing machine is cleaned and dried thoroughly.
- Los Angeles machine consists of a standard cylinder which is rotated for 500 or 1000 revolutions at a rate of 20 to 33 rpm.
- The sample of aggregates is placed in the cylinder and subjected to rotation.
- Particles smaller than 1.7 mm are separated and weighed.
- The loss in weight gives the abrasion value of aggregates.
- The abrasion value should not be more than 30% for pavement works or heavy wearing surface and 50% for low wearing surface.

2.1.3 Testing of Concrete

There are various tests carried out on concrete both on fresh concrete and hardened concrete. The two most common tests carried out are as follows:

(a) **Test for Workability for Fresh Concrete:** Following are the tests carried out to determine the workability of concrete:

(i) **Slump Test:** This is the simplest test carried out and this test can be carried on site at the point of placing the concrete. The details of this test are specified in IS7320. The test has following apparatus:

- Slump Cone: This is in shape of a frustum of a cone having the following dimensions. The cone is 10 cm in diameter at the top, 20 cm in diameter at the base and has a height of 30 cm.
- Tamping Rod: The tamping rod is a MS rod of 16 mm in diameter and 60 cm in height.

- Platform: A flat platform made of MS sheet to place the cone.

Following are the steps to carry out the test:

- Slump mould is placed in position over the platform.
- Fresh concrete is poured in cone in four equal layer.
- Each layer is compacted before the next layer is poured:
- Once the concrete fills the cone, the cone is removed slowly and the concrete is allowed to subside.
- The depth of the subsidence from the top is measured and is known as slump.
- The values of the slump for various concrete works are mentioned in the Table 6.1.

Table 6.1: Table showing the value of slump for various works

Type of work	Slump in mm
Road or Pavement	20 – 30
Beams and Slabs	50 – 100
Columns and Retaining walls	75 – 150
Canal Lining	70 – 80
Arches and Lintel	90 – 100
Mass Concreting	25 – 50
Pile Concrete	150 – 175

(ii) **Compaction Factor Test:** This test has been specified in IS1199. In this test the concrete to be tested is allowed to fall for a standard height which ensures that standard amount of work is done to compact the concrete.

The apparatus consist of two hopper vessels with trap doors at the bottom, and once cylinder to collect the sample. The apparatus and the arrangement are shown in figure.

Following are the steps to carry out the test:

- The apparatus is arranged as shown in figure.
- The sample of concrete is placed to the upper hopper to the brim of the hopper.

- The trap door is opened and the sample is allowed to fall to the lower hopper.
- The trap door of the lower hopper is then opened and the sample if allowed to fall into the cylinder.
- The excess concrete remaining above the top level of the cylinder is then cut off with the help of plane blades.
- The concrete in the cylinder is weighed. This is known as weight of partially compacted concrete (W_1).
- The cylinder is filled with a fresh sample of concrete and vibrated to obtain full compaction. The sample is placed in 5 layers and compacted properly.
- The concrete in the cylinder is weighed again. This weight is known as the weight of fully compacted concrete (W_2).
- Compaction factor is the ratio of partially compacted concrete to fully compacted concrete. C.F = W_1/W_2.
- The closer the value is to 1, the better is workability of concrete sampled.

(iii) **Vee-Bee Test:** This method is generally suitable for very dry concrete whose workability is low. This method is carried out in laboratory.

The apparatus consist of following parts and the arrangement is shown in figure:

- Vibrating table.
- Cylindrical Metal pot.
- Metal cone also known as slump cone.
- Tamping rod made of MS.

The test has the following procedure:

- Slump cone is placed inside the metal cylindrical pot.
- Slump test is carried out in the cylinder.
- Then the sample is vibrated mechanically and timed.

- Vibration is carried out till such time the sample loses its conical shape.
- The time taken to achieve the shape change is called the vee bee degree.
- Concrete is more workable if the vee bee degree is low.

(b) **Test for Compressive Strength for Hardened Concrete:** Concrete is good in compression and this is the first test carried out on hardened concrete. IS516 describes the testing procedure of concrete.

Following is the testing procedure:

- The samples of fresh concrete are prepared as specified in IS1199.
- Fresh concrete is then placed in a mould. The mould could be cube or a cylinder. The size of the moulds is specified in IS10086.
- For cubes, the mould size is 150 mm and for a cylinder the size is 150 mm in diameter and 300 mm in height.
- The concrete shall be placed in the mould in a layer of not more than 50 mm and compacted.
- The compaction can be done either by hand or manually. For hand compaction a tamping rod confirming to IS10086 is used and the number of strokes per layer is 35.
- The sample is the kept for 24 hours and then removed from the mould and cured in clean water.
- The concrete sample to be tested is to be fully saturated and surface dry concrete.
- The samples are generally tested for 7 days and 28 days. However the acceptance criteria are considered only for 28-day strength.
- The samples are tested in CTM (Compression Testing Machine) and tested for compression. Figure 6.5 shows a Compression testing machine.

Fig. 6.5: Figure showing Compression testing machine

2.1.4 Testing of Steel

Steel is an important part of RCC structures and it is necessary to ensure that the good quality steel is used in construction for strength and durability of structure. Following is the test conducted on steel to be used as reinforcement:

(a) **Reverse Bend Test:** Reverse Bend test is a common test carried out on steel and it is done on site. It is done to find out the ductility of steel. In this test the steel to be used is bent to an angle of 135° around a mandrel. The bent bar is then aged by keeping in boiling water for 30 mins and allowed to cool. The bar is then bent in reverse direction to an angle of 157.5°. There should be no fracture in the bent portion. This is specified in IS1786.

(b) **Tensile Test:** Steel used in RCC takes care of tensile loads and hence it needs to be checked for its tensile properties. The tensile properties are the tensile strength which is 0.2% of proof strength and percentage elongation. The procedure and details of these tests are specified in IS1608. The stresses are calculated using effective cross-sectional area.

2.2 Quality Control of Concrete

Quality control of concrete is the process where the quality of concrete is maintained during the production of concrete.

It should be ensured that the deviations from the prescribed standards are kept to the minimum.

2.2.1 Factors Affecting Quality of Concrete

The factors affecting the quality of concrete are as below:

(a) **Quality of Raw Materials:** It is important that the raw materials used for concrete adhere to the prescribed standards. Any variation in the quality of raw material will affect the quality of concrete directly.

(b) **Quality of Batching and Mixing Equipments:** Proper batching ensures that the concrete produced is of good quality and strength. The ratio of mix proportions need to be properly controlled especially in higher grade of concrete. Even a small variance will reduce the strength of concrete drastically and make it ineffective. If the equipments are faulty then the results will vary.

(c) **Quality of Workmanship During Production of Concrete:** Human errors can reduce the quality of output and desired results will not be achieved. It is necessary to reduce the human errors to the minimum. As far as possible, it is to be ensured that mechanical means are adopted for production of concrete. This will help in isolating the source of problems.

(d) **Transporting of Concrete:** Concrete needs to be transported and placed within a stipulated time. For every mix, the initial setting time is specified. It is also to be ensured that the quality of concrete is not affected during the transporting. The most common error which is done during the transporting of Ready mix concrete over large distance is the mixing of water by the driver of the transporting truck. This is done to ensure the workability of concrete in case of delay of transporting of concrete due to traffic. This should be totally avoided.

(e) **Placing and Compaction of Concrete:** Placing of concrete should be done as per procedure for each of the member. Placing of concrete varies for slab, beams, columns and walls and proper procedure should be followed for concrete of each of the members. Compaction is a very important procedure to ensure the density of concrete. Following of incorrect method of compaction can result in segregation of concrete.

(f) **Curing of Concrete:** Curing is the most important activity which will ensure the strength of concrete. It would not be an exaggeration to state that very good curing will cover some of the defects during the production of concrete. Improper curing will result in concrete not gaining its required strength.

2.2.2 Deterioration of Concrete

Deterioration of concrete is defined as the decay of concrete when it is under duress due to external and internal factors. The external factors can be natural or manmade. The natural factors can be attributed to oxidation due to moisture which can result in rusting of reinforcement. The man-made factors could be chemical attacks like sulphates, chlorides and salt solution. The internal factors could be attributed to alkali-aggregate reaction.

2.2.2.1 Causes of Deterioration

The main cause of deterioration of a concrete structure is due to corrosion. The corrosion of can take place either in concrete or in steel.

(a) **Corrosion of Concrete:** The corrosion of concrete is a gradual process which takes place over the years due to the attack of aggressive chemical agents. Generally concrete is not impervious and there are voids which will allow the atmosphere to pass through. This process will lead to chemicals infiltrating the concrete. Even the presence of cracks will allow and enhance such processes.

Generally infiltration of acids facilitates the formation of calcium salts when they react with free lime present in concrete. These salts will become soluble and start seeping out. Seeping out of inherent chemicals make the concrete weaker thus reducing the strength of concrete. This is also called leaching.

(b) **Corrosion of Steel:** Corrosion of steel is the process wherein the iron present in steel gets oxidised. This process is also called the rusting of steel. Following are the factors which influence the corrosion of steel:

- Insufficient cover to reinforcement.
- Void in concrete.

- Presence of chemical agent in concrete which facilitate oxidation.

There are mainly two mechanisms of corrosions:

- **Dry Corrosion:** In dry corrosion, oxidation of steel takes place due the exposure of steel to atmosphere and the oxygen present in air chemically reacts with iron in steel to form ferrous oxide.
- **Wet Corrosion:** In wet corrosion, oxidation of steel takes place due the exposure of steel to water and the oxygen present in water chemically reacts with iron in steel to form ferrous oxide.

2.3 Quality Control During Construction

It is not sufficient to ensure that the raw materials to comply with the quality standards. It is also very important to ensure that the quality during construction with respect to manpower and process so that the required finish is obtained. Any gaps in the process will not give the desired results on site. The workmanship is checked both before the work is carried out and after the work is completed. Following are the basic steps to ensure that the quality is maintained on site for various works. Depending on the type of work and the level of quality control more steps can be added on site.

2.3.1 Concreting

Maintaining a good workmanship in the concreting process is utmost important to ensure that the structure is strong and of good finish. Following are the points to be taken note of on site for concreting:

(a) Checking steps involved in pre-work:

- Check if the layout is as per drawing.
- Check if the starter is placed properly (in case of column and wall casting).
- Check if the formwork is fixed as per specification.
- Check if the formwork is prepared as per specification like cleaning, oiling etc.
- Check if the reinforcement is placed as per drawing.

(b) Checking steps involved in post-work checking:
- Check the level of top surface finish.
- Check for surface cracks.
- Check if the curing is done as per requirement.
- Check if the de-shuttering is done as per specification.
- Check for any irregularities in concrete workmanship like honeycombs and cracks.

2.3.2 Blockwork

Though in framed structures block work is a filler material and covers the structure, the quality of workmanship in blockwork is of utmost importance. Any lack in quality in workmanship of blockwork will have cascading effect on the subsequent finishing activities. Following are the points to be taken note of onsite while carrying out the blockwork:

(a) Checking steps involved in pre-work:
- Check the layout marking of block work including openings, electrical dropouts from ceiling and holds for equipment insertions.
- Check the right angle for wall.
- Check if the RCC surface is clean of unwanted projections.
- Check the quality of block.
- Check if the specification of mortar mix.

(b) Checking steps involved in post work.
- Check the openings of joineries.
- Check the line, level and plumb of the wall.
- Check the pointing of joints.
- Monitor curing.

2.3.3 Plastering

Maintaining a good workmanship in plastering is necessary to ensure a good finish. Though this is an aesthetic requirement and does not affect the strength in general, sometimes cracks in plastering can facilitate seepage which will make the structure weak over a period of time. Following are the points to be taken note of on site for plastering:

(a) Checking steps involved in pre-work:
- Check if the service lines are completed.
- Check if the electrical drop outs are exposed (for ceiling plastering).
- Check for line and plumb of the wall.
- Check if the surface is clean of unwanted projections.
- Check if the wire mesh is provided at joints.
- Check if the surface is wet.

(b) Checking steps involved in post-work:
- Check for finish of plaster.
- Check the levels and alignment.
- Check the internal room sizes.
- Monitor curing.

2.3.4 Placing of Joineries

Fixing of joineries is the final step in the covering of a building which will make it closed. Though it looks more of an aesthetic work, it can have effects on structural strength especially in case of load bearing walls.

(a) Checking steps involved in pre-work:
- Check if the openings are correct to plumb and level.
- Check for line and plumb of the wall.
- Check if the joineries are as per specifications.

(b) Checking steps involved in post-work:
- Check the alignment of frame both level and plumb.
- Check if the joineries are sealed as per specification.
- Check the cleaning work of panels.

2.3.5 Flooring

Laying of flooring is one of the two finishing works which is purely aesthetical. However sometime poorly laid flooring can allow the water to seep into the structure and cause problems in dampness. If allowed to persist for a longer time, water can seep into the concrete through the joints and facilitate deterioration.

(a) Checking steps involved in pre-work:
- Check if the material is as per approved colour, size, shade etc.
- Check if the level of finished flooring is marked as per design.
- Check if the location is clean and devoid of any projections.
- Check the proportion of mortar mix

(b) Checking steps involved in post-work:
- Check for the finish of floor visually.
- Check for the levels and slope of finished floor.
- Check if the packing and pointing of joints is done as per specification.
- Check if the surface is cleaned after completion for handover.

2.3.6 Painting

Painting is the second of the two finishing works which is purely aesthetical. Completion of painting gives a final finish to the building and a completed look. A badly painted surface can be an eyesore for a long time.

(a) Checking steps involved in pre-work:
- Check if the material is as per approved colour, shade etc.
- Check if the finish of plastering is proper (before application of wall putty).
- Check if the surface is smooth (After application of putty).
- Check if the location is clean and devoid of any projections.
- Check the proportion of mortar mix.

(b) Checking steps involved in post-work:
- Check for the finish of paint visually.
- Check for brush marks on the surface.
- Check the edges and corners are in line.

- Check for overlapping of paints (for double cloured walls).
- Check for any pin holes.
- Check if the surface is cleaned after completion for handover.

2.4 Repair and Rehabilitation

Any structure needs to be maintained regularly and protected against external factors to ensure the durability of structure and longitivity of service criteria. The structure can be placed under distress due to various factors. The factors which put the structure under duress are as follows:

- Distress due to heavy external load.
- Distress of concrete due to weathering of concrete.
- Distress due to poor construction practices.
- Distress due to chemical attack.
- Distress due to differential settlement.

The most common types of distresses seen in concrete are Honeycomb and Crack.

2.4.1 Honeycomb

By definition honeycomb means a structure which has cavities. In concrete, honeycombing is the term used to describe areas of the surface that are coarse and stony. Honeycombs can be simply defined as a rough, pitted surface or voids in concrete resulting from incomplete filling of the concrete against the formwork.

They are often caused by using concrete that is too stiff or by not vibrating it sufficiently after it has been poured. Normally honeycombs were observed in structural elements like Columns, lift walls and a few beams as well. Figure 6.6 shows honeycombs in various structural elements.

(a) **Causes of Honeycomb**: The causes of honeycomb and voids include the following reasons:
   - Concrete having low workability.
   - Excessive vibration of concrete.

- Improper placing practices such as pouring or releasing the concrete from a height.
- The normal water cement ratio that should there is within the range of 0.4 to 0.6, but if more water is been added the workability of the concrete mixture increases. This will cause problems when the vibrator is been inserted.
- The grout will start to leak through the small voids and from the bottom voids of the formwork.
- It will also cause the plywood to swell and will also change the dimensions and the cross-section area of the structural elements.

Fig. 6.6(a): Figure showing Honeycomb in column

Fig. 6.6(b): Figure showing Honeycomb in beams

Fig. 6.6: Figures showing honeycombs

(b) **Repair of Honeycombs:** Honeycombs whenever observed should be treated immediately. The steps involved in repair of honeycomb is as follow:

- Initially the concrete particles and loosened aggregates should be removed by using a chipping hammer and a wire brush.
- Next the area should be cleaned thoroughly with a wire brush to remove finer particles and should be washed.
- Let the surface dry well and apply "Chemi-fix" glue on the area.
- Next mix the construction grout along with white cement and water as recommend by the manufacturers specifications (the reason for adding whit cement is to get the required colour of the mixture). It's a large honeycomb as in figure 4.4 concrete mixture is been poured after making a pocket.
- Afterwards the mixture should be poured or pasted on the area where the honeycombs have occurred.
- The formwork should be removed after 12 hours and should be cured.

2.4.2 Cracks

Cracks are the openings observed in hardened concrete. Visibility of cracks indicates that the structure is under distress in some form various reasons. Some of the reasons are settlement of building due to settlement of soil, expansion and contraction due to temperature variance, wear and tear of materials, overloading etc. Whatever be the reason, it is important to understand the type of crack and the reason for its occurrence before taking any remedial action. Figure 6.7 shows some types of cracks.

2.4.2.1 Classification of Cracks

Cracks can be broadly classified under following category:

(a) **Classification Based on Strength:** Based on the affect on strength of the structure the crack is classified as below:

(i) **Structural Cracks:** Structural cracks are those which result from incorrect design, faulty construction or overloading and these may endanger the safety of a building and their inmates.

(ii) **Non-Structural Cracks:** Non-Structural cracks occur mostly due to internally induced stresses in building materials. These cracks normally do not endanger the safety but may look unsightly, create an impression of faulty work or give a feeling of instability.

(b) **Classification Based on Further Activity:** Based on the movement of crack, it can be classified as:

(i) **Active Cracks:** Active cracks are those types of cracks which show some change in direction, width or depth over a measured period of time.

(ii) **Dormant Cracks:** Dormant cracks remain unchanged in width or depth.

If left unrepaired, both active and dormant cracks provide channels for moisture penetration, which can lead to future damage.

(c) **Classification Based on Size:** Based on the size the crack can be classified as:

(i) **Micro Crack:** By definition, a micro crack broadly means that the crack can be observed only through microscope. Generally there cracks are not visible to naked eye. The negligence of micro crack can result in the crack turning into a macro crack.

(ii) **Macro Crack:** Macro cracks are those cracks which are visible to naked eye. These cracks have a tendency to have an effect on structural durability and need to be attendant immediately.

2.4.2.2 Types of Cracks

The various types of cracks and its possible causes are mentioned below:

(a) **Plastic-shrinkage Crack:** This type of crack runs to the mid-depth of the concrete and is distributed across the surface unevenly, and is usually short in length. This type of crack often occurs while concrete is setting, due to the surface of the concrete drying too rapidly relative to the concrete below.

(b) **Crazing/Map Cracking/Checking:** This type of crack is a formed as a web of fine, shallow cracks across the surface of the concrete. It generally occurs due the surface of concrete drying faster than the interior concrete. Since this is a surface crack it does not cause immediate structural distress.

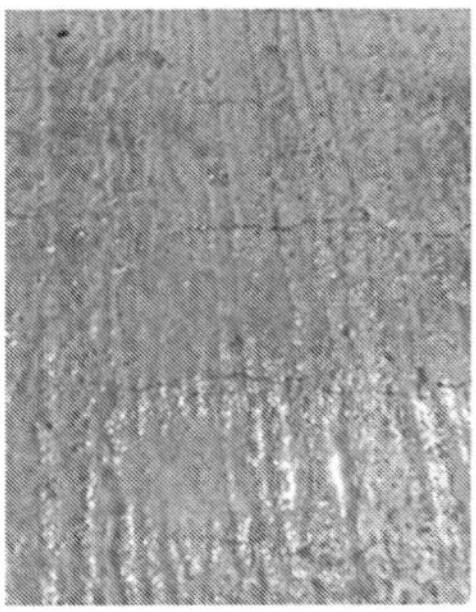

Fig. 6.7(a): Figure showing shrinkage cracks

Fig. 6.7(b): Figure showing calculations due to cracks

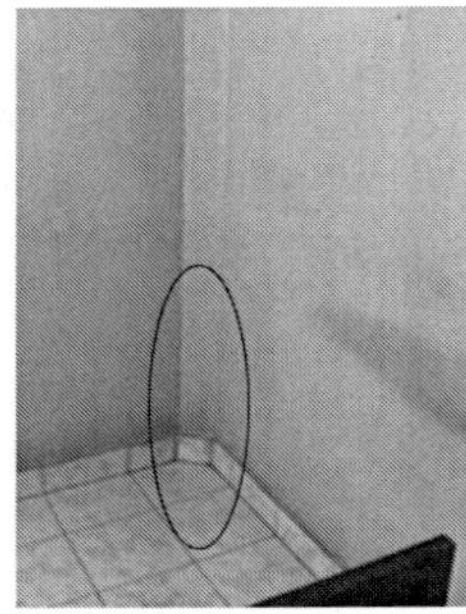

Fig. 6.7(c): Figure Showing crack due to settlement

Fig. 6.7: Figure showing various types of cracks

(c) **Hairline crack:** They are very thin but deep cracks which occur generally when the concrete is setting. They are minute and generally difficult to locate immediately. When neglected these cracks can expand to major crack.

(d) **Pop-Outs:** These are conical depressions in the concrete surface. They occur when a piece of aggregate near the concrete surface is particularly absorbent, causing it to expand and pop out of the surface of the concrete.

(e) **Scaling:** Small pock marks in the concrete surface, exposing aggregate underneath. Concrete is not impervious and sometimes during curing the water seeps through to the interiors of concrete. Water that seeps into the concrete will expand when it freezes, pushing off pieces of the concrete surface.

Scaling can also be caused by delamination, which occurs when too much water (due to insufficient curing) or air (due to insufficient vibrating) remains in the concrete when it is finished. The water and air rise to the top and form pockets below the surface. These pockets may form blisters or which may break open to create scaling.

(f) **Spalling:** Spalling is the surface depression that is larger and deeper than scaling, often linear along the length of reinforcement. Spalling is also caused by pressure from under the surface of the concrete. It occurs mostly due to improperly constructed joints or the corrosion of reinforcement in the concrete. Corrosion creates pressure as rust forms, which can push away large chunks of concrete, and expose the corroded metal below. Spalling that exposes corroded metal can be particularly problematic because the corrosion is likely to accelerate due to exposure to air and water.

(g) **D-Cracking:** These are cracks that runs roughly parallel or stem from a concrete joint and are deeper than surface cracks. Due to moisture infiltration at the joint.

(h) **Offset Cracking:** These are cracks which occur where the concrete on one side of the crack is lower than the concrete on the other side.

Due to uneven surfaces below the concrete, such as subgrade settlement or pressure from objects such as tree roots, previously-placed concrete, or reinforcement.

(i) **Diagonal Corner Cracking:** These are cracks which run diagonally i.e. generally at an angel of 45 from the corners. The diagonal cracks can occur both in concrete as well as brickwork.

The corners of concrete slabs can be prone to curling (due to differences in temperature at different depths in the curing concrete) or warping (due to differences in moisture evaporation at different depths in the curing concrete). The dryer or colder level of concrete will shrink more and create cracks as the concrete dries. Because the warped or curled-up corners often have some empty space below them, they are also prone to cracking after curing due to weight overload causing the corner to snap downward into the empty space.

The diagonal cracks can also occur in walls. This occurs due to the stress developed because of expansion of wooden joineries.

2.4.2.3 Evaluation and Analysis of Cracks

Evaluation of crack is essential to identify the probable cause of cracks. Proper evaluation will help in identifying the extent of distress and deciding the method of repair. The cracks can be analysed by two methods namely:

(a) **Visual Inspection:** In this method, the location and size of cracks are observed visually and analysed using a crack comparator. Generally a crack comparator can detect cracks up to a width of 250 microns. This is very useful to determine if the crack is an active crack or a dormant one. However the drawback of this is that it cannot detect cracks of very low width.

(b) **Non-destructive Testing:** With the improvement in technology, advanced equipments have been developed to determine the status and location of cracks. The most common testing is ultra sound testing where in even the depth of cracks can be determined by sending ultra sound pulses. This is an accurate method of testing.

2.4.2.4 Repair of Cracks

After the cracks are evaluated appropriate method of repair needs to be adopted to restore the structure to its original serviceable condition. There are two stages in the repair process:

(a) **Preparation of Surface:** The surface of the concrete which is to be treated has to be readied for application of the repair material. Any repair without proper preparation of surface will be inadequate. The steps adopted for the preparation of surface is as follows:

- Unsound material should be completely removed.
- Cracks should be removed from the surface and a rough surface should be created.
- Surface should be thoroughly cleaned and dried and free from any foreign material or latence.
- If the surface is oily then the fat should be removed by application of detergents like caustic soda or tri-sodium-phosphate.

(b) **Application of Repair Material:** The selection of material and its application is dependent on the type of crack and its extent of repair. The methods adopted for application of repair material is as follows:

(i) **Dry Packing:** This method is preferred for dormant crack and cracks of smaller sizes. In this method the crack is widened to about 25 mm width and 25 mm depth. Rich cement mortar (1:3) made of sand of less than 1.18 mm is placed. The water content of the mortar is also kept low to form a hard paste.

(ii) **Stitching:** It is generally adopted where the surface is under tension. The stitching procedure consists of drilling holes on both sides of the crack and placing of U-shaped MS bars or high tensile steel and grouting them with either a non-shrink grout or epoxy resin.

(iii) **Grouting:** In this method the cracks are pumped with thin cement mortar are injected into the surface which fills up the voids and gaps. This method is adopted when the cracks are deeper. It is an expensive method.

(iv) **Shotcreting or Guniting:** Guniting is the most effective process of repairing the concrete surfaces. In this process, the cement mortar is applied on the surface to repaired using a gun powered by a pneumatic pump at high velocity. This process can also be adopted to improve permeability.

Reference and Further Reads

Li, Zongin, "Advanced concrete technology", 2011, John Wiley & Sons inc.

Rumane, Abdul Razzak, "Quality Management in Construction Project", 2010, CRC Press.

Shetty, M.S., "Concrete Technology Theory and Practice", 1982, S. Chand Publishing.

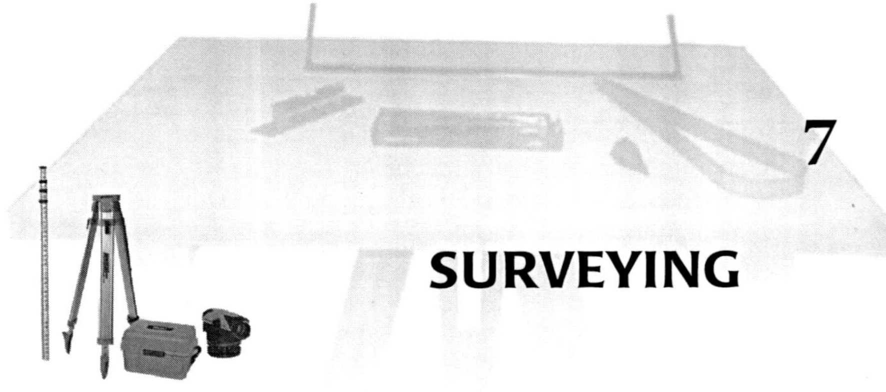

7

SURVEYING

Surveying is the art of determining the relative position of points on, above or below the surface of the earth by means of direct or indirect measurements of distance, directions and elevations. Surveying can be one dimensional, two-dimensional or three-dimensional depending on the requirement. Surveying can be as simple a process as finding a distance between two points to a highly complex one of determining the volume of space.

Basic surveying has been a part of construction from the early civilization. The principle of survey was based on simple geometry. The ancient structures built during the early civilizations indicate their command over surveying. The earliest recorded recognition of surveying as a profession is from the roman civilization. The Romans created a register of their lands especially conquered lands for levy of taxes.

The survey of India was established in 1767 by the British East India Company to assess its territory. It is one of the oldest engineering departments of the Indian Government. The first survey was conducted for the region of baramahal (present day Dharmapuri, Krishnagiri and North Arcot) in 1793. The height of Mt. Everest was estimated for the first time in the year 1892.

1. DIVISION OF SURVEYING

The surveying is primarily divided into two namely plane surveying and geodetic surveying.

1.1 Plane Surveying

It is that type of surveying in which the mean surface of the earth is considered as plane surface and curvature is not taken into account. This type of survey is effective only for small area upto 260 km^2 or 100 sq miles. The degree of accuracy is less.

1.2 Geodetic Surveying

It is that type or surveying in which the curvature of the earth is considered. This is also known as trignometrical surveying. All the maps are prepared using this type of survey.

It is defined as a survey of portion of country measuring a single base and connecting it with various points in the tract surveyed by a series of triangles, the angles of which are carefully measured.

The great trignometrical survey was a project which was embarked on in 1802 by an infantry officer of East India Company. The purpose of the survey was to measure the entire Indian subcontinent with precision.

2. CLASSIFICATION OF SURVEYING

Surveying can be classified based on the following criteria: Classification based on the nature of field of survey, Classification based on object of survey, Classification based on method of survey, Classification based on instruments use.

2.1 Classification Based on Nature of Field of Survey

Based on the field or region of survey, surveying is classified into the following:

2.1.1 Land Survey

As the name suggests, such a type of survey is conducted to determine the natural features on land such as layout, extent and contour of land, layout of water bodies such as rivers, natural canals, lakes, streams and layout of flora, i.e. forest lands.

2.1.2 Marine Survey

As the name suggest, such a type of survey is conducted mainly on large water bodies for the purpose of navigation, construction of marine or offshore structures such as harbours, oil rigs and also to determine the mean sea level.

2.1.3 Astronomical Survey

Such a kind or survey is conducted to determine the layout and observations of heavenly bodies, solar systems and galaxies. In this type of survey, a general mapping of a region of sky is done which lacks a specific observational target. There is a lot of significance attributed to the astronomical survey in olden days. The orientation of structures was based on astronomy and the relative position and motion of various celestial objects. This is predominantly used in development of celestial almanacs. On a lighter note, it is very useful for astrologers for their predictions.

2.2 Classification Based on Object

Based on the object or reason for survey, surveying is classified into the following:

2.2.1 Engineering Survey

A survey undertaken for the purpose of obtaining the information for planning of an engineering project is known as engineering survey. The most common engineering project is planning of a town or a township.

2.2.2 Military Survey

Military survey is carried out to obtain information, by visual observation or other detection methods, about the activities and resources of an enemy or potential enemy, or about the meteorological, hydrographic, or geographic characteristics of a particular area, from the point of view of defence and battle.

2.2.3 Mining Survey

Mining survey is carried out to explore the mineral wealth in a particular location. Mining surveying is generally defined as the art of making such measurements, as may be necessary,

- To determine the location and extent of bodies of minerals like coal, mineral ore, etc.
- To determine the relative positions of points in the mine with regard to each other or to points on the surface.

2.2.4 Archeological Survey

In archaeology, survey or field survey is a type of field research by which archaeologists (often landscape archaeologists) search for archaeological sites and collect information about the location, distribution and organization of past human cultures across a large area (e.g. typically in excess of one hectare, and often spanning many km^2). Archaeologists conduct surveys to search for particular archaeological sites or kinds of sites, to detect patterns in the distribution of material culture over regions, to make generalizations or test hypotheses about past cultures, and to assess the risks that development projects will have adverse impacts on archaeological heritage. The surveys may be:

(a) Intrusive or non-intrusive, depending on the needs of the survey team and the risk of destroying archaeological evidence if intrusive methods are used; and

(b) Extensive or intensive, depending on the types of research questions being asked of the landscape in question.

Surveys can be a practical way to decide whether or not to carry out an excavation (as a way of recording the basic details of a possible site), but may also be ends in themselves, as they produce important information about past human activities in a regional context.

2.2.5 Geological Survey

Geological survey is conducted to determine the different strata of earth crust. It is the systematic investigation of geology beneath the given piece of ground for creating a geological map or model. One of the major used of this survey is the information on orientation and measurement of bedding planes and intersection faults. This is very helpful in determining plate movement and more importantly in predicting earthquakes in a region.

2.3 Classification Based on Method Used

Based on the method used for surveying there are two types of surveying:

2.3.1 Traverse Survey

Traversing is that type of survey in which a number of connected survey lines form the framework and the directions and lengths of the survey lines are measured with the help of an angle measuring instrument and a tape or chain respectively. There are mainly two types of traversing:

(a) **Closed Traversing:** In closed traversing the lines form a circuit which ends at a start point. It is in shape of a closed loop. This type of surveying is suitable for surveying boundaries of objects and for large areas.

(b) **Open Traversing:** As the name suggests, in open traversing the ending point of lines do not come back to the starting point. This type of surveying is suitable for long narrow strips of land like coast.

2.3.2 Triangulation Survey

Triangulation is a surveying method that measures the angles in a triangle formed by three survey control points. Using trigonometry and the measured length of just one side, the other distances in the triangle are calculated. This method is very popular in making maps and also in navigation.

2.4 Classification Based on Instrument Used

There are various types of instruments used for surveying. These instruments can range from basic instruments like a tape to highly advanced instruments like total station. Based on the type of instrument used the surveying is classified as below:

2.4.1 Chain Survey

Chain survey is the type of surveying where the measurement and plotting is done only by measuring lengths. It is the simplest form of survey. The first formalized chain was introduced by Edmund Gunter in 1620. Gunter's chain consists of 100 links, each link measuring one foot. Gunter's chain was later replaced by metal tapes for linear measurements when the system of SI units was introduced. Following points to be taken care while conducting a chain survey:

- The number of survey lines should be kept the least possible number required.
- Survey lines should form well conditioned triangles.
- Tie lines should be provided to avoid long offsets.
- Chain lines should lie on the ground.

- Care should be taken to ensure the straightness of the lines.
- The length should be measured accurately.

(a) **Instruments Used in Chain Survey:** Following are the instruments used in chain survey:

  (i) **30 m Chain:** This is the main instrument which defines this type of survey. Chain is a metal chain, connected by links. 100 links make a chain. So effectively the 30 m chain has 100 links, one link has a length of 0.3 mts which is roughly one foot.

  (ii) **Cross Staff:** Cross staff is an instrument used to set out right angles. It consists of a either a frame or a box with two pairs of vertical slits mounted on a pole. There are mainly two types of cross staffs namely open cross staff and French cross staff.

  (iii) **Ranging Rod:** It is a wooden peg with a cast iron shoe which is driven into ground. It is painted with alternate colours of white and red for easy sighting. This is used for marking boundaries as well as objects. The object vane of alidade is oriented towards the ranging rod.

  (iv) **Arrows:** Arrows are pegs used to mark the boundaries and distances.

  (v) **Tape:** A tape is used to measure the fraction distance which cannot be measured by the chain.

(b) **Terminology Used in Chaining:** Following are the terminology used in chaining:

  (i) **Offset:** When there are objects in between the line of chaining, then the measurement is done by going around the object. The points are transferred to a parallel location and the chaining is continued. This process is called offsetting and the line joining the main line and the parallel line is called offset. Offsets having lengths less than 15 m are short offset and more than 15 m are long offset. When the offset is at an angle, then it is called oblique offset.

(ii) **Random Line:** During linear measurement, if the end station is not visible from any intermediate point, then the random line method is used. In this method, a random line is drawn from the end station to a point which is visible from both the end station and the intermediate station.

(iii) **Base Line:** The longest of the main survey line is called the base line. All the survey lines are plotted with reference to the base line. It is the most important line and it should be plotted accurately. Any error in the base line will result in a totally wrong survey.

(iv) **Well Conditioned Triangle:** Triangle which is very close to the shape of an equilateral triangle is called a well conditioned triangle. In a well conditioned triangle, all the angles should be in the range of 30 to 120 degrees.

2.4.2 Leveling

Leveling is the simplest form of survey for measuring of heights. It helps in determining the levels of the surface and preparation of topographic map.

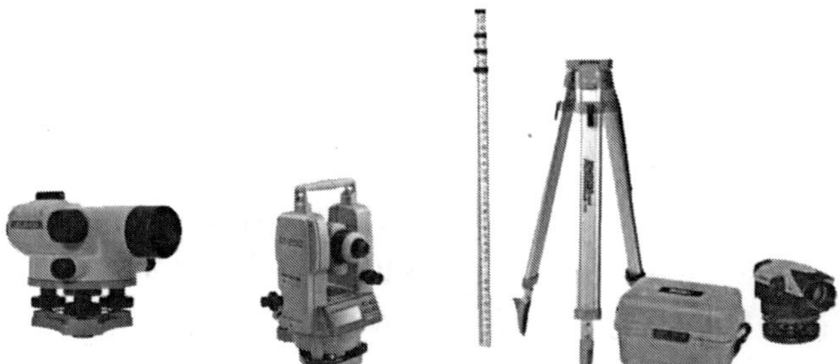

Fig. 7.1: Figure showing a leveling instrument

(a) **Methods of Leveling:** There are mainly three methods of leveling namely:

(i) **Spirit Leveling:** It is the simplest and most accurate form of leveling. This is a direct method of leveling in which the vertical distances are determined by drawing of horizontal lines and calculating the relative difference in heights.

(ii) **Trigonometric Leveling:** This is an indirect method of leveling where in the elevation of points are computed from vertical angles and horizontal distances.

(iii) **Barometric Leveling:** This type of leveling uses the pressure difference between two points to calculate the height. The assumption in this is that the pressure varies linearly as the height. This is an approximate method and used where it is not possible the access the location.

(b) **Leveling Instrument:** Leveling instrument has the following parts:

- **Level:** A level is the instrument which is used to sight the level. The most common leveling instrument is the dumpy level.
- **Leveling Staff:** The leveling staff is rectangular member which is calibrated. The foot of the staff denotes zero level.

2.4.3 Plane Table Survey

Plane tabling is a graphical method of surveying in which field observation and plotting are done simultaneously. It is generally used when the surveying area is small and accuracy required is not very detailed. It is a graphical method of survey. As the name suggests this type of surveying is done using a plane table. The origin of the concept of plane table is not clear. The earliest mention of plane table is found in Abel Foullon's "Usage et description de l'holomètre", published in Paris in 1551.

(a) **Instruments used in plane table surveying:** The various instruments used in this type of survey are as below:

(i) **Plane Table:** Plane table consists of a well seasoned board made of good quality wood such a teak mounted on a tripod. The whole system is called a plane table as it represents a form of table. The sizes of the board vary from 40 cm × 30 cm to 75 cm × 60 cm. There are three types of table based on the method of adjustment of tripod and precision namely:

- **Traverse Table:** In this type of the table, the tripod is adjusted manually and precision is estimated visually.
- **Jhonson's Table:** In this type of table, the size of the board is 75 cm × 60 cm. The head of the tripod contains a ball and socket joint. A spirit level is used to level the table.
- **Coast Survey Table:** This type of table has a very high precision.

(ii) **Alidade for Sighting:** It generally consists of a wooden or metal rule with two vanes at the ends. The vanes are hinged in such a way that they fold down on the rule when not in use. The metal used in alidade is either brass or gun metal. The length of the rule is 50 cm. The leveled edge of the alidade is called the fiducial edge. One of the vanes is provided with a narrow slit while the other vane is open and contains a thin wire. Both the slits provide a definite line of sight which can be aligned to the object to be sighted. The alidade can be rotated so that the alidade is in a straight line with line of sight of the object and line can be drawn against the working edge of the alidade. There are two types of alidade:

- **Plain Alidade:** This is generally used where the location to be surveyed is closer to the base. Plain alidade is shown in figure 7.2.
- **Telescopic Alidade:** This is used to survey locations farther to the base as the telescope mounted on the alidade can view locations at a distance.

(iii) Plumbing Fork: Plumbing for or U frame as it is known consists of a U with arms of equal length and a plumb bob which is suspended from the lower arm. The plumbing fork is used to centre the plane table over the station over the known point which has already been plotted.

(iv) **Spirit Level:** This is used to level the plane table. The ground on which the set up is fixed can be uneven. Spirit level ensures that the plane table is on true level.

(v) **Compass:** Compass or trough as it is called is a kind of ruler which is used to orient the table to magnetic north.

(vi) **Ranging Rod:** It is the same as described in chain survey.

Fig. 7.2: Figure showing plane table

Fig. 7.3: Figure showing compass

(vii) **Drawing Sheet with Rain Proof Cover:** Since the survey is captured on site, the drawing sheet should be of good quality and it should not get affected by climatic conditions. Rain proof cover is kept ready to protect the drawing sheet from getting wet while the survey is being carried out.

(b) **Advantages of Plane Table Survey**: Following are the advantages of plane table survey:

- There is no possibility of missing out measurements.
- The actual survey is transferred to paper as it is done on site.
- It is the most suitable method for preparing a small map.
- The process is fast and inexpensive.
- It is useful in areas where there is magnetic interference.
- Errors can be easily detected and mistakes occurring from transferring measurements from field book are avoided.

(c) **Disadvantages of Plane Table Survey**: Following are the disadvantages of plane table survey:

- The accuracy of work is limited.
- It is not possible to carry out this type of survey during wet climate.
- Transporting of equipments is cumbersome and missing out of accessories is a possibility.

2.4.4 Compass Survey

Compass surveying is a type of surveying in which the directions of surveying lines are determined with a magnetic compass, and the length of the surveying lines are measured with a tape or chain or laser range finder. The compass is generally used to run a traverse line and this process is known as compass traversing.

The earth acts as a powerful magnet and the earth's magnetic field generally runs from south to north. The compass does not measure the angel between the two lines but measure the angle between the line and the magnetic meridian.

There are two types of compass used in this type of survey:

(a) Prismatic compass

(b) Surveyors compass.

The difference between prismatic compass and surveyor's compass is shown in Table 7.1:

Table 7.1: Table showing the difference between prismatic and surveyor's compass

S. No.	**Prismatic**	**Surveyors**
1	The graduation circle is fixed to broad needle and does not rotate with the line of sight	The graduation circle is fixed to the box and rotates with the line of sight
2	There is a prism at the viewing end	There is a slit at the viewing end
3	The graduations are in WBC system	The graduations are in Q.B. system
4	The graduations are marked inverted	The graduations are marked directly
5	Magnetic needle does not act as index	Magnetic needle acts as index
6	Instrument can be hand held	The instrument needs a tripod

2.4.5 Tacheometer Survey

Tacheometry is a branch of angular surveying in which the horizontal and vertical distances are determined using the relative angles between the points without the use of chain or tape. It is derived from the Greek word which means quick measure and true to its name it is a system of rapid surveying. This type of surveying is very useful in difficult terrain like valleys, steep grounds, water streams etc. There are two methods of tacheometric survey namely.

(a) Stadia Method

(b) Tangential Method.

2.4.6 Theodolite Survey

Theodilite is an instrument used to measure angles in horizontal and vertical planes. The general principle in theodolite survey is traversing. A theodolite may be either transit or non-transit. In a transit theodolite, the telescope can be inverted in the vertical plane, whereas the rotation in the same plane is restricted to a semi-circle in a non-transit theodolite.

A modern theodolite consists of a movable telescope mounted within two perpendicular axes namely the horizontal axis or trunnion axis and zenith axis. When the telescope is pointed at an object, the angles of each of the axis can be measured with great precision.

Though various versions of theodolites have been used, the invention of first modern theodolite is credited to Jesse Ramsden in 1787. Edward Samuel Ritchie invented the waterborne version of theodolite using a pendulum device to counteract wave movement.

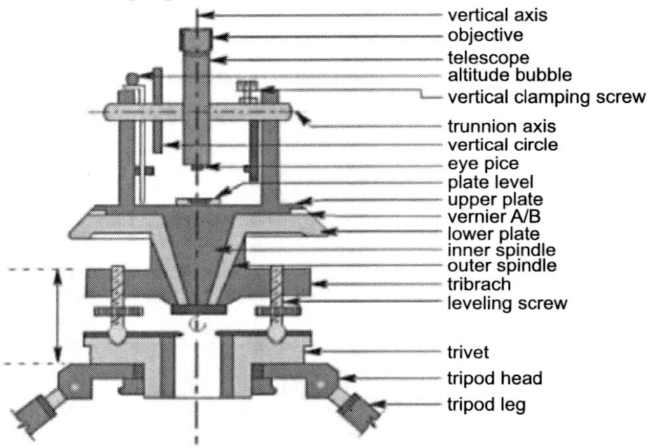

Fig. 7.4: Figure showing theodolite

2.4.7 Total Station Survey

Total station is the most advanced instrument used in surveying. Fundamentally this can be taken as a combination of two instruments namely electronic distance measuring instrument and electronic theodolite. The development of total station has changed the face of surveying. With one instrument angles, distance and coordinates can be measured and the data can be recorded and stored internally and reproduced when required.

Total station has undergone major developments from its early invention in 1970's. The robotic total station eliminates the role of operation. It can be operated by a remote control. GPS total station has integrated GIS and real-time kinematic concept.

Like any other methods the total station survey has advantages and disadvantages.

The advantages are:

- Surveying is quick, efficient and accurate.
- Multiple surveys can be performed from one location.
- Data can be plotted easily by uploading the recorded data.

The disadvantages are:

- The accuracy of vertical elevation is lesser than the conventional techniques.
- The horizontal coordinates are calculated based on rectangular system rather than spheroid system.

Figure 7.5 shows a schematic diagram of a total station. Figure 7.6 shows a sample output from a total station survey.

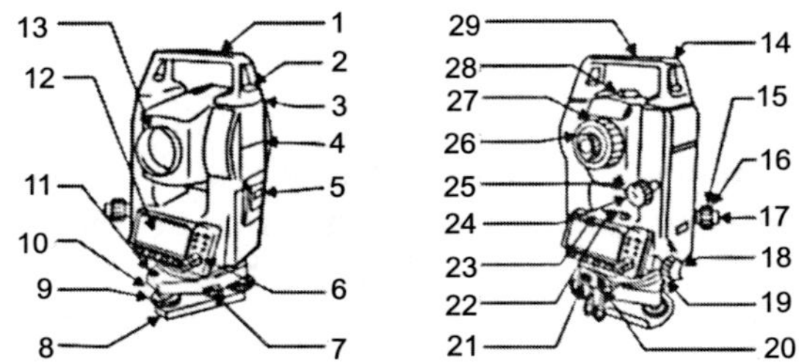

1. Handle
2. Handle securing screw
3. Data input/output terminal (Remove handle to view)
4. Instrument height mark
5. Battery cover
6. Operation panel
7. Tribrach clamp
8. Base plate
9. Levelling foot screw
10. Circular level adjusting screws
11. Circular level
12. Display
13. Objective lens
14. Tubular compass slot
15. Optical plummet focusing ring
16. Optical plummet reticle cover
17. Optical plummet eyepiece
18. Horizontal clamp
19. Horizontal fine motion screw
20. Data input/output connector
21. External power source connector
22. Plate level
23. Plate level adjusting screw
24. Vertical clamp
25. Vertical fine motion screw
26. Telescope eyepiece
27. Telescope focusing ring
28. Peep sight
29. Instrument center mark

Fig. 7.5: Figure showing total station

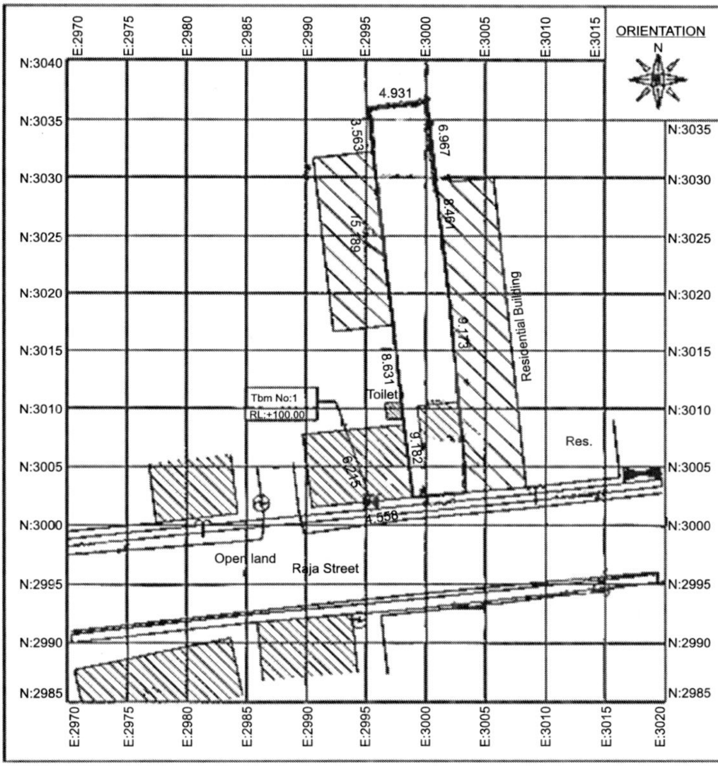

Fig. 7.6: Figure showing output of a total station survey

Reference and Further Reads

Anderson, James M. and Edward M. Mikhail, "Surveying—Theory and Practice", 2013, 7th Edition, McGraw Hill.

Punmia, B.C., Arun K. Jain, Ashok K. Jain, "Surveying – Vol 1", 2016, 17th Edition, Laxmi Publication.

Roy, S.K., "Fundamentals of Surveying", 2011, 2nd Edition, PHI.

8 TRANSPORTATION

Over the year the modes and speeds of travel from one place to another has ensured that this branch gains prominence as a specialisation. From the early days when human kind travelled on foot and no proper pathway was necessary to the present day of high speed trains, expressways and supersonic air travel, the field has evolved by leaps and bounds.

The mode of transport is categorized based on the freedom of movement the path for the transportation provides. There are three modes of transport based on the degree of freedom of movement:

- Transport with one degree of freedom—Railway
- Transport with two degree of freedom—Roadways and Waterways
- Transport with three degree of freedom—Airway.

1. RAILWAY

Railway is a form of transportation with one degree of freedom, which means that the movement is only in one direction. The reason why it is defined to have one degree of freedom that it moves on a fixed path called rails which does not allow any lateral movement.

1.1 History

The wagon ways are sometimes considered the predecessor of railways. Wheeled vehicles were pulled by animals and were used for transporting heavy goods. This concept was later extended to tramways which were used for transporting passengers. Invention of steam engine by James Watt paved way for the replacement of animal drawing carriages to mechanically powered carriages. William Murdoch produced a working model of self-propelled steam carriage.

The first full scale steam locomotive was built by Richard Trevithik in 1804. Inspired by his predecessors, George Stephenson built the first flanged wheel locomotive in 1814. Subsequently in 1825, he built the first railway which was opened to public on 27th September 1825 in UK. The pioneer to run the first locomotive was George Stephenson.

In India the first railway run was on 16th April 1853 between Mumbai (then Bombay) and Thane for a distance of 33 kms. Over a period of time, railways have become the most preferred mode of transport due to its low cost. Indian Railway is the largest employer in the world.

1.2 Rail Gauge

Rail gauge is the clear distance between the two rails for a single track take at the flange. When the railways started, the gauge was a standard gauge to match with the horse drawn carriages for hauling of cargos in coal mines. George Stephenson used 4 ft 8.5 in (1435 mm) as the gauge which was half an inch more to cater for free movement at binding of curves. Eventually this came to be known as standard gauge and is predominantly used in most countries today.

(a) **Types of Gauges:** There are mainly four types of gauges:

(i) **Broad Gauge:** The size of broad gauge varies in different countries. In India, the value is 1676 mm (5 ft 6 in) and also called the Indian gauge. This gauge is predominantly in Indian subcontinent. Some of the other variance in size of broad gauge used in different countries is the Russian 5 ft and Irish 5 ft 3 in.

(ii) **Standard Gauge:** Originally used by George Stephenson, this is used as the universal gauge in most countries. The value of the gauge is 1435 mm with a tolerance of 10 mm either side.

(iii) **Meter Gauge:** As the name suggests, the value of the gauge is 1000 mm, though technically it is a form of narrow gauge. This is used generally in metros in most cities and in some places in India. In India, all the meter gauge lines are being converted to broad gauge for uniformity.

(iv) **Narrow Gauge:** Originally any gauge below the standard gauge was termed as narrow gauge. Most of the narrow gauges are between 2 ft (610 mm) to 3 ft 3 in (1000 mm), though sometimes the meter gauge is also considered narrow gauge.

(b) **Factors Governing the Selection of Gauge:** Following are the factors which govern the selection of gauge:

(i) **Return of Investment:** This is one of the major factors governing the selection of gauge. For a broader gauge, the cost of construction is higher but if the traffic is not sufficient to offset the cost and workings don't indicate a profitable return, then appropriate gauge should be adopted for maximizing the returns.

Fig. 8.1: Figure showing parts of permanent way

(ii) **Speed:** Generally the speed of the train is dependent on the wheel diameter and the wheel diameter is adopted as 0.75 times the gauge. Hence the speed of the train is dependent on the gauge. For higher speed of trains, it is preferable to adopt a higher gauge.

(iii) **Topography:** Higher gauges have larger radius of curvature. Mountainous terrains have sharper turns and curves. Under such circumstances, it is advisable to go for narrower gauges.

1.3 Permanent Way

Permanent way is the system which is created for the train to move. It is the system which is made of various components. Each component has a specific role to play. Figure 8.1 shows the section of permanent way. There are three components of permanent way namely Rail, Sleeper and Ballast.

1.3.1 Rail

Rail is a continuous steel girder supported by sleepers. It is subjected to downward moving load due to movement of train.

(a) **Function of Rail:** Rail has following functions:
- It provides smooth surface for the movement of train.
- It provides hard surface to withstand the load of moving train.
- It transfers load coming from train to sleepers safely.

(b) **Ideal Requirements of Rail:** The rail should ideally have the following requirements:
- The top surface of the rail should be smooth and even so that the friction is minimal.
- The top surface must be hard enough to withstand the crushing.
- Bending compression and tension stress should be same.
- It should transfer load safely to sleeper.
- Corrosion of section should not take place.
- Composition of material must be uniform.
- Profile should be convenient for working and fixing.

(c) **Chemical Composition of Material used for Rail:** The chemical composition of the steel used for rail is shown in Table 8.1.

Table 8.1: Table showing the chemical composition of rail

S.No.	Element	IRS T18	IRS T12
1	Carbon	0.40 to 0.65%	0.60 to 0.80%
2	Manganese	0.80 to 1.45%	0.80 to 1.30%
3	Silicon	0.03 to 0.30%	0.10 to 0.50%
4	Sulphur	0.060%	0.035%
5	Phosphorus	0.060%	0.035%

(d) **Rail Section:** There are various types of rail sections used. The most common ones are:

(i) Double Headed

(ii) Bull Headed

(iii) Flat Footed.

1.3.2 Sleeper

Sleeper is the second constituent of permanent way. It is the transverse member which supports rail and transfer the load coming on the rail to the ballast section. It is laid perpendicular to the rail section.

(a) **Functions of Sleeper:** The sleeper has following functions:

- It supports the rail.
- It transfers the load from rail to ballast.
- It ensure that the gauge is maintained correctly.
- It absorbs shock experienced by the rail from movement of train.
- It acts as an insulator for the rails.

(b) **Materials used for Sleeper:** From the early days when stone blocks were used as sleepers, the development of sleepers has come a long way. Originally the rails were placed just over the sleeper which caused a problem in slip. Eventually a system of bolt and nuts were developed to connect the rails to the sleepers. Following are the various materials used for sleeper:

(i) **Wooden or Timber:** This type of sleeper was the earliest form of sleeper used in permanent way due to its availability in nature. However before using it, it needs to be seasoned properly and coated to ensure that decaying is prevented. Though the life cycle of timber sleeper is short, the lifespan being 15-20 years, the advantages of being easily available, good insulation and easier to work with has made it the most widely used sleeper material. The dimensions of the sleeper used for various gauges are as follow:

Table 8.2: Table showing the sleeper sizes for various gauges

Gauge	Length (m)	Width (m)	Thickness (m)	Base Area (m^2)
Broad	2.47	2.54	0.1250	0.465
Meter	1.83	2.03	0.1125	0.310
Narrow	1.52	1.78	0.1125	0.209

The suitability and strength of timber to be used as sleeper is determined by a value known is composite sleeper index or CSI. The formula for CSI is I = (S +10H)/20 where

S is the strength of timber at 12% moisture content

H is the hardness index of timber at 12% moisture content

According to Indian railway board, the standard value of CSI is as below:

- For sleepers used for track = 783
- For sleepers used at crossing = 1352
- For sleepers used at bridges = 1455

(ii) **Steel:** With the reduction in availability of timber and its ever growing cost, it became necessary to look at alternative material for timber. The commercialization of Bessemer's process made steel a viable material to replace timber in construction. Steel sleeper's advantage over timber was that the load carrying capacity and life span was more and it was suitable for trains to ply at a higher speed.

(iii) **Cast Iron:** Fixing of bull headed rail was difficult to the steel sleepers. To easy out, cast iron sleepers were adopted instead of steel sleepers. The section of cast iron sleeper is shown in figure. The cast iron sleeper consists of a semi-circular hollow pot and the two sleepers in a row connected by a tie rod to maintain the gauge. The diameter of the pot is 50cm at the base and the cross section of tie rod is 51 × 13 mm. The advantage of this type of sleeper is its long life and high scrap value. The disadvantage is that it corrodes easily and requires high maintenance.

(iv) **Precast:** The manufacture of concrete and the discovery of reinforcing it with steel made this type of sleeper the widely used in today's world. It has a very good lifespan and it compensates all the drawbacks of the previously used sleeper materials, namely timber, steel and cast iron. The discovery of pre-stressing made precast sleepers even more durable than the conventional RCC sleepers. Due to its heavy self weight, it is most suitable for high speed trains. The only drawback is the handling due to heavy weight and low scrap value.

Table 8.3 shows the comparison of various sleepers for different properties.

Table 8.3: Table showing the comparison of various sleeper materials

	Timber	Cast Iron	Steel	Precast
Weight	Light	Medium	Medium	Heavy
Life (years)	12 – 15	35 – 40	35 – 50	40 – 60
Handling	Easy	Tedious	Easy	Difficult
Maintenance	More	Less	More	Least
Insulation	Excellent	Unsuitable	Unsuitable	Good
Packing of Ballast	Easy	Difficult	Easy	Easy
Spikes	Simple	Complicated	Simple	Simple
Corrosion Resistance	No Corrosion	Minimum	Minimum	Maximum
Suitability to High Speed	Less	More	More	Ideal
Scrap Value	Nil	High	High	Less

(a) **Sleeper Density:** Sleeper density is the number of sleepers used for one rail length. Sleeper density is dependent on the following factors:

- **Gauge:** Generally the broader gauges will have higher sleeper density. The empirical formula for sleeper density is taken as n + x, where n is the length of rail in meters and x is the constant which is dependent on gauge. The value of x is 6 or 7 for broad gauge, 4 or 5 for meter gauge, and 3 for narrow gauge.

- **Axle Load:** Load coming on sleeper also determines the density. The load carrying capacity of sleeper is fixed. The higher the axle load, the sleeper density will be more.
- **Speed of Train:** Similar to axel load, higher speed of train will transmit higher dynamic load and hence resulting in higher sleeper density.
- **Type of Ballast:** If the strength of ballast material is more, then the spacing of sleeper is less.
- **Nature of Embankment:** Formation level is the final level where the load is going to be transferred. A weaker embankment will result in closely placed sleeper.

1.3.3 Ballast

Ballast is the third and the final part of the permanent way. It supports the sleeper and completes the system of permanent way. It is generally made up of coarse material and of a considerably large size. The dimensions of ballast section for various types of gauges are shown in Table 8.4.

Table 8.4: Table showing the dimension of ballast section for various gauges

Gauge	Top width	Depth	Vol. per mt run
Broad	3.35 m	20 – 25 cm	1.036 m^3
Meter	2.25 m	15 – 20 cm	0.750 m^3
Narrow	1.83 m	15 cm	0.460 m^3

The slope is generally maintained in the range of 1:1 to 1:5.

(a) **Functions of Ballast:** Following are the functions of ballast:

- It transfers the load coming on the sleeper to the sub-soil.
- It provides elasticity to the tracks.
- It defines the total width of permanent way.
- It checks the growth of vegetation.

(b) **Ideal Requirements of Ballast:** The material used for ballast should have the following requirements:

- It should have a very good crushing strength.
- It should be chemically inactive.

- It should not absorb or retain water or moisture.
- It should be a bad conductor of electricity.

(c) **Materials Used for Ballast:** The various material used for ballast are as follows:

(i) **Broken Stones or Gravel:** This is the most commonly used material. Gravel can be found on river bed and can be used naturally. However the problem with gravel is that it has low friction and lesser load carrying capacity.

Broken stones or coarse aggregates are products taken from breaking granite. Granite is an igneous rock and very high load carrying capacity. The size of the broken stones should range from 19 mm to 51 mm and it should be graded for good density of ballast. The size of stone ships adopted for various types of sleepers are as follows:

- For Timber sleeper – 51 mm
- For Steel sleeper – 38.5 mm
- For CI sleeper – 19 mm

(ii) **Broken Bricks:** Broken granite can be expensive and to reduce the cost broken bricks can be used. These can be obtained from over-burnt bricks which are unsuitable for construction. These materials have good drainage properties. However the drawback of this material is that they have less crushing strength and can be crushed into powder under continuous cyclic loading.

(iii) **Burnt Coal:** Burnt coal was obtained from coal used to propel steam engines. Earlier days, the only type of engine used was steam engines and steam engines used coal for generating energy. Burnt coal was available in plenty and freely available. It was the cheapest material to be used for ballast. Another advantage was that during floods, when procuring of broken granite or bricks are difficult, this is easily available and also has good drainage properties. However, like broken bricks, they can be easily crushed, have less durability and can corrode the sleepers and rail.

(iv) **Sand:** Sand is a natural material and found in abundance on the river beds. Even though the material is cheap and easily available and has good draining properties, it has major drawbacks which make it difficult to be used as ballast. Due to its light weight, it can be displaced by wind. It will also be difficult to maintain the sections in position when the train speed is high. The finer particles of sand may enter into reciprocating parts of locomotive and increase the friction and wear and tear. The chemicals in the sand can cause corrugation of rail.

(v) **Furnace Slag:** Slag is the byproduct obtained from manufacturing of steel. This has similar properties of sand and can be used in case of emergencies when other materials are not available.

(d) **Screening of Ballast:** Due to continuous movement of the train, the ballast may be displaced from the original position. There is also a possibility of some amount of ballast being crushed under the loading. This will result in reducing the quantity and density of ballast which will affect the stability of tracks. Crushing of granite may also fill up the voids which will hamper the draining properties of ballast. To maintain the required properties of ballast, screening of ballast is carried out.

Screening of ballast is the process by which the smaller particles and the dust particles are removed from the mix. This is done by a sieve. A wire mesh of size 1.5 m × 1.2 m, made up of required sieve is placed in inclined position. The ballast to be screened is sieved by forcefully throwing the ballast on the mesh. The ballast retained on the sieve is reused. Generally two labourers are required to screen the quantity of ballast for the one rail length.

1.4 Stations and Yard

Stations and yards, in simple terms, are places for resting and interchanging. One is for passenger and other is for trains.

1.4.1 Station

A station is a place on a railway line where the train stops to load or unload passengers or freight. The part which connects the train to the station is called platform. A station can have one platform or number of platforms. Largest station in the world is Grand central terminal with 44 platforms. In India Howrah station has 23 platforms. The longest platform is in Gorakhpur station which incidentally is the longest platform in the world too. The station where the train reaches the end of the line is called a terminus.

The selection of site for a station has the following criteria:

- It should be near a city or a town and well connected by road.
- It should have enough space for parking and future expansion.
- It should be avoided where the line curves.

1.4.2 Yard

A railways yard is a series of tracks for storing, sorting, loading or unloading of trains. A rail yard has many tracks in parallel for keeping rolling stock stored off mainline so that the flow of traffic is not obstructed. It is equivalent to a garage for rolling stock. Broadly there are three types of yard based on the topography of slope and the principle of shunting. The different types of yards are hump yard, gravity yard and flat yard.

Fig. 8.2(a): Showing railway station

Fig. 8.2(b): Showing railway yard

Fig. 8.2: Figure showing railway station and railway yard

1.5 Rolling Stock

Rolling stock is defined as that part of the train which moves in rails. It is so called because the components of rolling stock literally roll. The rolling stock is also considered a liquid or movable asset since the value of the vehicle can be readily estimated. There are three types of rolling stock:

1.5.1 Locomotive

Locomotive or engine, as it is more popularly called, is the part which converts chemical or electrical energy to mechanical energy to ensure movement of the train. It is the power generator for the train.

(a) **Types of Engine:** Based on the type of energy used as source, the engine is categorized into the following:

   (i) **Steam Engine:** In this type of engine, the steam is generated by heating water using coal as the medium of fuel and the steam generated is used to power the train. Steam engines were the original form of locomotive. Coal was used to create steam and it had harmful effects on the environment. In 1985 the steam engines were phased out of Indian railways.

   (ii) **Diesel Engine:** In this type of engine, diesel is used as a medium of fuel to power the engine. The world's first diesel powered locomotive was operated in 1912 after the patent of diesel engine by Rudolf Diesel. Constant improvements, increase of hauling capacity and a cleaner fuel resulted in diesel locomotives replacing steam engines from 1950's.

(iii) **Electric Engine:** In this type of energy, electrical energy is converted to mechanical energy to power the engine. First known electric locomotive was built by Robert Davidson in 1837 using galvanized cells. The first electric passenger train was presented by Werner Von Siemens in 1879. It consisted of 3 cars and was powered by DC engine. Invention of Alternating current by Nichola Tesla made it possible to electrify the railway network over long distances.

(b) **Design of Locomotive:** Design of locomotive mainly deals with the size of locomotive and the wheel design. There are two different classification of wheels namely,

(i) **Idle Wheel:** The front and the rear wheels are also known as idle wheel and their role is to distribute the weight of the engine.

(ii) **Driving Wheel:** The intermediate wheel is also known as the driving wheel and it is directly connected to the piston. The diameter of the intermediate wheel is larger than the front and the rear wheel.

1.5.2 Coach

Coaches are that part of rolling stock which carry passenger. They are also known as passenger coach vehicle. In earlier days, the coaches had minimal facilities with seating arrangement for passengers. As the technology developed, more and more facilities were introduced and the coaches were categorized into classes based on the facility provided in the coach. Development of air-conditioning system ensured that the coaches could be air-conditioned. Improvement of speed of the train and introduction of high-speed trains needed the coaches to be entirely covered to reduce friction to train and discomfort to passengers. This made air-conditioning necessary for the entire train.

(a) **Classes in Coach:** The various classes in coach in India are as follows:

(i) **Third Class:** Third class was the basic form of coach used in earlier days especially before independence. The seats were wooden benches and there were three tiers of berth. Berth is the type of seat or platform where there passengers can lie down in full stretch comfortably.

(ii) **Second Class:** In second class the facilities were better than third class in the sense that the seats were more comfortable. The number of tiers of berth continues to be three.

(iii) **First Class:** First class became the symbol of luxury of travel. In this class the seats were wider and the number of tiers of berth on each side was two. Furthermore the coach was divided into cabins and each cabin had four berths only which gave more privacy.

(iv) **Air-Conditioned Class:** With the advancement in the concept of portable air conditioning, this class came into vogue in India in 1980's. For major trains, one or two coaches were air-conditioned. Initially the air conditioned class was two tiered. To make it more economical and the preference of passenger of air conditioned class over first class, three tier air-conditioned class was introduced.

(v) **Chair Car:** Increase in speed and reduction in time travel between destinations made day journey possible in train over long distances. Chair car is a type of class where the passengers sit for the entire journey. There is no berth. The chair car is both air-conditioned as well as non air-conditioned.

1.5.3 Wagon

Wagons are that part of rolling stock which is used to transport goods. They are also known as other coach vehicles to differentiate it from passenger coaches. Wagons can be covered or uncovered depending on the type of goods they are carrying. Wagon is defined by the quantity and type of load they are carrying.

1.6 Special Railways

The conventional railway system consists of a combination of locomotive and wagons travelling on a pair of rails. The length of the train is long and it is economical for long distances. Due to the need for economical intra city public transport, shorter railways were developed. Further advancements in technology have also seen the development of trains not touching the rail. The day is not far when the world will see trains running without rails.

1.6.1 Light Railway or Tram

Trams are rail vehicles which are used for intra-urban public transportation. They are sometimes referred to as cable car. It is called light railway because of the capacity it carries. It generally has two coaches and runs on electric energy. The very first tram was operated on Swansea and mumbles railway in UK. It was horse drawn street car, a successor to stage coach. Electric trams were invented and tested by Fyodor Pirotsky in Saint Petersberg, Russia in late 19th century. In India, horse drawn trams were introduced in Kolkata (then Calcutta) in 1873. This was upgraded to electric tram in 1900, though the first electric tram service was started in Chennai in 1895.

Fig. 8.3(a): Conventional Railway

Fig. 8.3(b): Tram

Fig. 8.3(c): Monorail

Fig. 8.3(d): Maglev

Fig. 8.3: Figure showing types of railway system

1.6.2 Monorail

The permanent way of a railway consists of two tracks. Mono rail is a type of railway which the permanent way has only one track. Monorails are generally elevated in nature and sometimes called as sky bus. The structure for a monorail consists of a single beam which acts as both the support and a guide.

Monorails have been experimented on from the early 19th century, seeking an alternative to conventional railways. The earliest form of mono rails were on ground. In the late 19th and early 20th century, the concept of suspended monorails was developed. The oldest surviving suspended version is the Wuppertal Monorail in Germany.

Mid 20th century saw a dip in the concept of monorail as it was perceived to be a high cost un-proven technology. From 1980 onwards, the popularity of monorail, especially in Japan picked up. With the cities becoming denser and lack of availability of space for expansion of roadways for increasing density of automobiles, monorail became a viable alternative for intra-city transportation.

1.6.3 Maglev

Maglev is a form of rapid transportation and is considered the future of railways. Maglev is derived from the word **mag**netic **lev**itation which is a method to move vehicles without making contact with the ground. Maglev can run on either mono rail or conventional double rail. The two types of technology used for Maglev are Electromagnetic suspension and Electrodynamics suspension which uses superconductivity.

Maglev can travel at very high speed (highest recorded speed—603 kmph) as it eliminates friction between the rail and train. The advantages include higher efficiency, lighter trains, less maintenance to name a few.

2. HIGHWAY

A road is a thoroughfare, route, or way on land between two places that has been paved or otherwise improved to allow travel by foot or some form of conveyance, including a motor vehicle, cart, bicycle, or horse. It has 2 degrees of freedom of movement in the sense that horizontally a vehicle can move in 2 directions.

The Organization for Economic Co-operation and Development (OECD) defines a road as "a line of communication (travelled way) using a stabilized base other than rails or air strips open to public traffic, primarily for the use of road motor vehicles running on their own wheels, which includes bridges, tunnels, supporting structures, junctions, crossings, interchanges, and toll roads, but not cycle paths".

2.1 History

The exact history of when the roads started is somewhat of an ambiguity. It is estimated that around by around 10000BC rough roads or pathways were used by human travelers.

The earliest travelers travelled on animals namely horses, oxen, donkeys or even elephants and did not need constructed tracks in open country. The first improvements of formation of roads were made by clearing trees and stone boulders and having a level surface.

Introduction of wheel and wheeled carriages made it necessary to pave the roads. Roads were paved with stone and bricks around 4000 BC and evidence of this are seen in Harappan civilization. The grand trunk road was built by Sher Shah Suri in 1540 connecting India from east to west. History suggests that parts of this road existed during the Mauryan Empire and Sher Shah Suri rebuilt it by extending it from Sonargaon near Dhaka in present day Bangladesh to Peshawar in present day Pakistan. The advent of 19th century saw new methods of highway construction due to the pioneering work of two British engineers, Thomas Telford and John Loudon McAdam, and by the French road engineer Pierre-Marie-Jerome Trésaguet

Pierre-Marie-Jerome Trésaguet is widely considered the pioneer in the scientific approach to road building. His memorandum on road construction in 1775 became the general practice in France. His method involved placing of a layer of smaller gravel over a bed of large rocks. The lower layer was known as the base course or the sub-base. The upper running course provided a smooth surface for movement and the lower layer transferred the load to the soil safely while protecting the ground from deformation due to movement of traffic.

Trésaguet's road design consisted of three layers of stones laid on a crowned sub-grade with side ditches for drainage. The first two layers consisted of angular hand-broken aggregate, maximum size 3 inches (7.6 cm), to a depth of about 8 inches (20 cm). The third layer was about 2 inches (5 cm) thick with a maximum aggregate size of 1 inch (2.5 cm). Top surface was given a smoother shape to ensure protection of larger stones in the road structure from iron of wheels and horse hooves. The drainage problem was solved by digging deep side ditches, making the surface as solid as possible, and constructing the road with a difference in elevation (height) between the two edges. The difference in height is known as the road's camber or cross slope.

Thomas Telford improved Trésaguet's on theory by introducing high quality stones and cubical stone blocks. Telford used partially shaped paving stones (pitchers), with a slight flat face on the bottom surface. He turned the other faces more vertically than Trésaguet's method. The longer edges were arranged perpendicular to the traffic direction, and the joints were broken like conventional brickwork. The smaller faces of the pitcher formed the upper and lower surfaces.

Broken stones were wedged into the spaces between the tapered perpendicular faces to provide the layer with good lateral control. Telford kept the natural formation level and manually cambered the upper surface of the blocks. He placed a 6 inches (15 cm) layer of stone no bigger than 6 cm (2.4 in) on top of the rock foundation. To finish the road surface he covered the stones with a mixture of gravel and broken stone. This structure came to be known as "Telford pitching". Telford's road depended on a resistant structure to prevent water from collecting and corroding the strength of the pavement. Telford raised the pavement structure above ground level whenever possible. Where the structure could not be raised, Telford drained the area surrounding the roadside. Previous road builders in Britain ignored drainage problems and Telford's rediscovery of these principles was a major contribution to road construction.

John Loudon McAdam, a Scottish Engineer is considered the pioneer in design of modern roads. He developed an inexpensive paving material made of soil and stone aggregate known as macadam. His method was simpler, yet more effective at protecting roadways. His method propagated that the native soil alone would be sufficient support the road and traffic upon it, as long as it was covered by a road crust that would protect the soil underneath from water and wear.

Size of stones used was the most important factor in McAdam's road building theory. The lower 200 mm (7.9 in) road thickness was restricted to stones no larger than 75 mm (3.0 in). The upper 50 mm (2.0 in) layer of stones was limited to 20 mm (0.79 in) size. A workman could check the stone size himself by seeing if the stone would fit into his mouth. The importance of the 20 mm stone size was that the stones needed to be much smaller than the 100 mm width of the iron carriage tyres that travelled on the road. Keeping the surface stones smaller than the tyre width made a good running surface for traffic. The small surface stones also provided low stress on the road, so long as it could be kept reasonably dry.

He also asserted that the quality of the road would depend on how carefully the stones were spread on the surface. Nothing was laid on the clean stone to bind the road. The action of the road traffic would cause the broken stone to combine with its own angles, merging into a level, solid surface that would withstand weather or traffic. The use of water absorbent substances was avoided. Mcadam laid his roads as level as possible. His 30-foot-wide (9.1 m) road required only a rise of 3 inches (76 mm) from the edges to the centre. Cambering and elevation of the road above the water table enabled rainwater to run off into ditches on either side.

2.2 Classification or Hierarchy of Roads

The general classification of roads throughout the world is more or less standard except for some minor differences in terms used. In India the roads are described in detail in Indian Road Congress (IRC). According to IRC106, Roads can be classified as the following:

(a) **Expressway:** It is the highest form of highway classification. It is also known as freeway or controlled access highway. This type of highway provides unhindered flow of traffic with no traffic signal, intersections or property access. The entrances and exits to the expressway are provided by slip roads. The elimination of conflicts with other direction of traffic improves the safety and capacity.

(b) **Highway:** The definition of highway somewhat lies in ambiguity of interpretation. By definition highway is a passage or a way over where in all the members using it can pass and re-pass without any hindrance. With the introduction of expressway, the term highway has undergone modifications. In general highway is type of road where any vehicular traffic can access.

(c) **Arterial Road:** Arterial road is a high capacity urban road and its main function is to deliver traffic from collector road to highways or freeways. It is a continuous route and it is the principal network in intra-urban travel. Arterials should be coordinated with existing and proposed highways for a seamless and efficient traffic collection and distribution.

(d) **Sub-arterial Road:** It is similar to arterial road but with a lower capacity.

(e) **Collector Road:** The role of collector road is to collect traffic from local road and feed it to arterial or sub-arterial road and vice versa. Normally full access is allowed on these roads from the abutting properties and they have very few parking restrictions.

(f) **Local Street:** Local streets are those types of streets which provide access to properties directly. These roads do not carry large volume of traffic. In this type of road, parking and pedestrian movement is unrestricted.

2.3 Parts of Road

The various parts of road are shown in Fig. 8.4 and described as below:

(a) **Carriageway:** This is the part of the roadway on which the traffic moves. The carriageway is described in number of lanes. The number of lanes varies from single lane road to 6 lane roads. Express ways can be design to 8 lanes too. The width of the carriage way for single lane is 3.5 m.

(b) **Central Median:** This part of the road divides the road between up traffic and down traffic. This helps is preventing access of vehicles from the opposite direction and enhances safety. As a general rule highways of 6 lanes and above should be provided with medians. For roads of 4 lanes, the provision of median should judicious, taking into various other factors. The width of the median is determined by various factors. As far as possible the width of the median should be uniform throughout.

(c) **Service Lane:** Service lane is local road running parallel to a high speed highway or expressway. It is also known as a frontage road. Frontage roads provide access to homes and businesses which would otherwise be cut off by a limited-access road and connect these locations with roads which have direct access to the main roadway.

(d) **Cycle Track:** Cycle tracks are dedicated lane provided for cycles and prohibits motorized vehicle. The minimum width of a cycle track is 2 m. In an effort to reduce pollution, people are encouraged to use cycle in many countries.

(e) **Kerb:** Kerb is the raised edge between the carriageway and pavement. The roads in urban locale are provided with kerbs. The kerb can be barrier type, semi barrier type or mountable type depending on the condition of roadway.

(f) **Footpath:** The footpath is the way provided for pedestrians. The minimum width of footpath as prescribed by IRC is 1.5 m. The design and construction of footpath should consider the safety of pedestrians and the volume of pedestrian traffic. The width of the footpath required should be calculated deducting all the hindrances like fences, tree, ramps etc.

(g) **Verge:** Verge is the narrow strip of land provided to accommodate lighting poles, supports for traffic lights and signs and underground services. It is located between the carriage way and the boundary of road. Verge is generally paved.

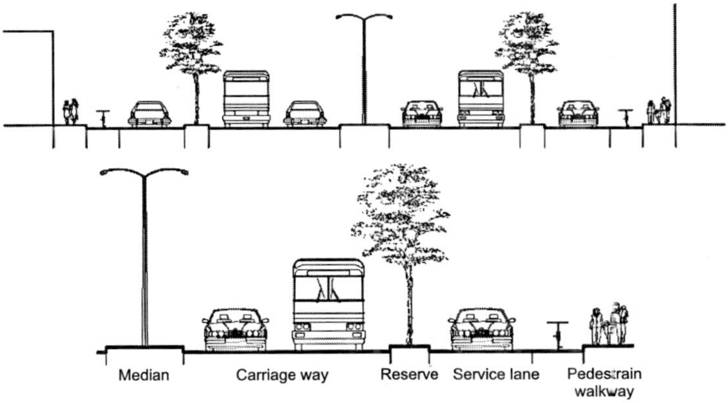

Fig. 8.4: Figure showing parts of road

2.4 *Specification of Bituminous Road*

The laying of road remain more or less the same suggested by Mcadam. The specifications for the laying of road are as below:

(a) **Sub-grade:** The sub-grade acts as a cushion for all the other layers of the road. Sub-grade is prepared by excavating the road to the required depth and compacting it thoroughly. The surface is moisturized by sprinkling water to enable better consolidation. Low spots developed during compaction are made good by additional materials. In rocky regions the sub-grades are not rolled whereas in region of clay soils, a layer of boulders packed with gravel or moorum is provided over natural ground. This is called soling. Nowadays, the soling is replaced by wet mix.

(b) **Sub-base:** Sub-base is prepared by spreading a mixture of 75 mm coarse aggregate and moorum or red gravel (depending on the availability) over the well compacted sub-grade. Sub-base is liad in either single layer or double layer. The number of layers is dependent on the soil condition and the density of traffic.

(c) **Base Course:** Base course is prepared by spreading a mixture of 50 mm coarse aggregate and moorum or red gravel (depending on the availability) over the well compacted sub-grade. Both the base course and sub-base should be compacted properly. The sub-base and base course together is termed as WBM (water bound mcadam). As is the case with soling, nowadays the WBM is replaced by bituminous mcadam where the red gravel is replaced by asphalt.

(d) **Surface Course or Wearing Course:** Surface course is prepared by spreading a mixture of 12 mm coarse aggregate and asphalt over the well compacted base (WBM). In some areas, to achieve smoother surface, an additional layer of premixed carper is laid. This is prepared by mixing 6 mm coarse aggregate and asphalt.

Fig. 8.5(a): Figure showing preparation of sub-grade

Fig. 8.5(b): Figure showing laying of WBM

Fig. 8.5: Figure showing laying of road

3. WATERWAY

A waterway is a mode of transport which has two degrees of freedom like roads. Any navigable body of water can be considered a waterway. Waterways can include oceans, seas, rivers, lakes and canals. Waterway must meet the following criteria:

- The waterway must be deep enough to allow the draft depth of the vessels using it.
- The waterway must be wide enough to allow passage for the beam width of the vessels using it.
- The waterway must be free of barriers to navigation such as waterfalls and rapid, or have a way around them such as canal locks or boats.
- The current of the waterway must be mild enough to allow vessels to make headway.

In the earlier days, waterways were used in its natural form for navigation and harbours but development of technology and need to shorten travelling distances paved way for human-made canals and artificial harbours. Suez Canal and Panama Canal are two examples of achievements by human race in the field of waterway.

3.1 Canal

Canals are human made channels build for both water supply and navigation. Earliest human made canals were constructed for irrigation purpose. There are evidences of extensive use of canals for the irrigation systems in the ancient civilization like Mesopotamian, Indus valley and Egyptian. Indus valley civilization had sophisticated irrigation and storage system. The reservoir built in Girnar in 3000 BC is one such example. Canal building for commercial use became popular in the middle ages. In the medieval age, water transport was cheaper than the land transport especially for transport of goods.

Canals built for water supply can either connect two rivers and supply water to intermediate areas or connect a water body like a river or lake to intermediate areas where there is no source of water for irrigation. For example, Indira Gandhi Canal was built to supply water to arid areas of Rajasthan from the rivers of Punjab.

Canals built for navigation generally connect two larger water bodies like sea or ocean and primarily built to shorten the time and distance of going around a land mass. For example, Suez Canal was built to connect Mediterranean sea and Indian ocean. In India, Buckingham Canal was built on the east coast connecting Vijayawada in Andhra to Villupuram district in Tamil Nadu.

Fig. 8.6(a): Showing waterway

Fig. 8.6(b): Showing navigable canal

Fig. 8.6(c): Figure showing sea port
Fig. 8.6: Figure showing seaport and canal

3.2 Port

Port is an artificial harbor which is built along the coast or shore where ships can dock and transfer people or cargo to and from land. In short port is a structure which connects land to water and vice versa. Port location is selected to optimize access to land and navigable water.

Ports are broadly classified as river ports and sea ports. River ports are also known as inland ports.

Design of port is a highly complex process and it requires wide range of competencies. Design of port must take into consideration the location, climatic conditions, type and volume of traffic to be handled to name a few. Since the port is exposed to seawater which is saline, factors like durability and corrosion of both concrete and steel have to be taken care of.

4. AIRWAY

Humankind's inherent ambition to fly like birds has been a dream from the day of evolution. The history of aviation goes beyond two thousand years ago with the earliest form of instruments of flying being kites and balloons. Leonardo Da Vinci's dream of flight found expression in several rational but unscientific designs, though he did not attempt to construct any of them. Discovery of hydrogen in 18th century made manned flights in balloons and airships a reality. Sir George Cayley's theory of aerodynamics laid the foundation for the principles of modern day aircrafts. He is also called the father of aeroplane.

4.1 Aerodrome

An aerodrome or airdrome is any place from where flight operations can take place. This can include an airstrip in the middle of nowhere to a commercial airport in the biggest city in the world. The various types of aerodrome are as follows:

(a) **Airstrip:** An airstrip is a small aerodrome which consists of just a landing strip to enable aircrafts to take off and land. They are generally on remote areas and areas where the population is low and the air traffic is minimal.

(b) **Air Base:** Air base is generally referred to an aerodrome which is used for defence purposes. It is also referred to as military air base.

(c) **Water Aerodrome:** A water aerodrome is an open area of water which allows aircrafts like seaplanes or amphibious aircrafts to take off or land.

(d) **Airport:** An airport is a special type of aerodrome that has satisfied certain certification criteria or regulatory requirements of the ICAO that a regular aerodrome may not have achieved.

4.2 Airport

As mentioned above airport is a type of aerodrome built to certain specifications. With air travel becoming more and more popular, the number of airports is on the rise. Design and construction of airports has become a very specialized branch. An airport has 3 parts namely,

(a) **Runway:** According to International civil aviation organization, a runway is defined as a rectangular piece of land prepared for landing and takeoff of aircraft. Runways may be artificial or natural. The artificial runways are made of either asphalt or concrete or a mixture of both. The design of runway depends on the ground conditions and the weight and volume of traffic it has to cater for.

Airport pavement construction is very expensive. The manufacturers of aircraft aim to minimize the stresses on the pavement. Manufacturers of the larger planes design landing gear so that the weight of the plane is supported on larger and more numerous tires. Attention is also paid to the characteristics of the landing gear itself, so that adverse effects on the pavement are minimized. To increase durability and load carrying capacity post-tensioned concrete are being used for construction of pavements.

Runway pavement surface is prepared and maintained to maximize friction for wheel braking. To minimize hydroplaning following heavy rain, the pavement surface is usually grooved so that the surface water film flows into the grooves and the peaks between grooves will still be in contact with the aircraft tires.

(b) **Terminal:** An airport terminal is a building where the passenger transfer takes place and includes facilities that allow the passengers to board or disembark an aircraft. The buildings that provide access to the aircraft are called the concourse. However the words concourse and terminal are used interchangeably.

Fig. 8.7(a): Showing Airport pavement and terminal

Fig. 8.7(b): Showing Aircraft hanger

8.7: Figure showing Airport

A satellite terminal is a building detached from other airport buildings, so that aircraft can park around its entire circumference. The first airport to use a satellite terminal was London Gatwick Airport.

(c) **Hangar:** A hangar is a closed building structure to hold aircraft in a protective storage. It simple terms it is a garage for aircrafts. Hangars are used to protect the airport from weather, direct sunlight and other natural forces. They are also used to carry-out maintenance and repair works of aircraft.

Reference and Further Reads

Chakroborty, Partha and Animesh Das, "Principles of Transportation Engieering", 2003, PHI Learning Pvt. Ltd.

Huang, Yang H., "Pavement Analysis and Design", 2003, Pearson.

Khisty, C. Jotin and B. Kent Lall, "Transportation Engineering: An Introduction", 2002, Pearson.

9 TALL STRUCTURES

Tall structures have fascinated human mind from the early civilization. Initially the constructions of tall structures were limited to defense purposes where the watch towers were built on the periphery of the fortress or townships. Subsequently the tall structures were adopted for various uses other than just for military. In early years there were many attempts to construct tall structures but have failed due to lack of technology and material. But in recent years improvement in technology, development of new construction material and knowledge of behavior of materials have helped in constructing building as high as 1 kilometer.

Tall is more of a relative term and there have been various definitions accorded to what can be construed for a tall structure. The perception of tall building also varies from place to place. In a city like Chicago or Hong Kong, a 15-storied build may not be considered so tall as compared to a 15-storied building in Chennai or Bengaluru. Various criteria have been taken into consideration to define a tall building.

On a general term a structure is defined as a tall structure when the lateral forces on the structure are more predominant than the gravitational forces. Structurally, when the height to width ratio is greater than 2 then also the structure is classified as tall structure.

The council of tall building and urban habitat (CTBUH) defines a tall building as one in which its tallness strongly influences its planning, design or use. It was formed in 1969 at Illinois. Its stated mission is to study and report on all aspects of planning, design and construction of tall buildings. The council maintains and ranks the height of tall structures. According to the council, a building above 300 mts is considered a super tall structure and above 600 mts is called a mega tall structure.

According to ASHRAE (American Society of Heating, Refrigerating, and Air Conditioning Engineers) technical committee for tall buildings, a building is defined as tall building if the height exceeds 91 m.

1. HISTORY

The initial concept of tall buildings came about from the need of business activities to be as close to each other and city center as possible, and limited available land space. The rapid growth of the urban population, high cost of land, the desire to avoid a continuous urban sprawl and the need to preserve important agricultural production have all contributed to drive of construction of buildings upwards.

The development of high rise buildings closely follows the growth of a city. The process of urbanization is a continuous one with people moving from rural areas to urban areas thereby forcing an increase in the density of the city. The primary concern was placing the maximum number of people on a minimum area of land. The resulting congestion and its impact on people and city as an organic interaction system was hardly a design consideration. The need for light, air, open ground level for public activity spaces led to evolution of the freestanding skyscrapers. In some cities like Honkong, Singapore etc. continuous growth of population concentration and local topography restriction make tall buildings the only feasible solution for housing needs.

One of the major advantages of tall structure is that it isolates different operations at different levels. For example one can have shopping complex in the lower stories, offices in the middle portion and residences at the upper stories. We can have an entire floor for services.

Tall buildings and structures are also built as prestige symbols as they form distinctive landmarks which symbolize the challenge to human effort and capability. The invention of elevators has also contributed to the growth of tall buildings.

The idea of an entire locality or a city in a single building is not far behind. Concepts are being developed and thought of, for an entire city in a single building so that limited land space is used for construction and there is lot of greenery around. This is a very important idea in present day due to the increase of global warming which is the direct result of reduction of green area. This will also reduce the pollution due to vehicle as all the function will be in the same building and the only mode of transport would be the elevator within the building.

The definition of building and tower are separated based on the percentage of occupiable space. According to CTBHU, to be considered a building, a structure should have atleast 50% of its height as occupiable. The structures which do not fall under this criterion are termed as tower.

High rise buildings can be put into following uses:

- Commercial uses such as office, store, bank, public utility.
- Residential.
- Industrial.
- Institutional.
- Public assembly.
- Multiuse.

Table 8.1: Table showing chronological increase in height of structure

Building name	Location	Year	Ht in mts	Material	Use
Pyramids	Giza, Egypt	2500 B.C.	147	Stone	Tomb
Lincoln Cathedral	UK	1311	160	Steel	Office
Washington Monument	Washington DC, USA	1884	169	Steel	Office
Eiffel Tower	France	1889	300	Steel	Office
Chrysler Building	New York, USA	1929	319	Steel	Office
Empire state building	New York, USA	1931	381	Steel	Office
Ostankino Tower	Moscow, Russia	1967	540	Steel	Office
CN Tower	Toronto, Canada	1975	553	Steel	Office
Burj Khalifa	Dubai, UAE	2007	830	Mixed	Multi purpose

2. DESIGN CRITERIA

The design of tall buildings whether for single use such as apartments, offices, schools, hospitals or for large scale multiple uses outlined above requires a team approach between various disciplines of design, material fabrication and construction. The approach to the design of tall building should be as a total system in which the physical support structure as an organic part grows with the design of the building. The structure cannot be considered separately as an unrelated addition to be plugged in into functional space later. The planning of the various systems should be given considerable thought. The uniqueness of each problem regarding a tall structure makes it a challenging job for an engineer.

The design criteria that apply to the design of tall buildings are:

2.1 Loading

The structure must be designed to resist the gravitational and lateral forces, both permanent and transient, which will be experienced both during its construction and subsequent service life. The building must cope up with vertical forces of gravity and the horizontal forces of wind above the ground and the seismic forces below the ground. The building envelope has to accommodate the differences in temperature, air pressure, and humidity between the exterior and the interior environments. These forces will depend on the size and shape of the building as well as on its geographical location. Maximum probable values for the forces must be established before the designing the structure. The structural elements must respond to all these forces within the limits prescribed.

2.2 Strength and Stability

It is important that the structure should be designed to have adequate strength to resist and remain stable under the worst possible load action that may occur during the lifetime of the structure. Forces and stress should be analysed for the most critical load combination, including augmented moments arising from II order additional deflection. The design should ensure that the structure does not slide or topple under lateral forces. The design should also include any additional stress caused by restrained differential moments due to special loads such as creep, shrinkage or temperature.

2.3 Stiffness and Drift

The provision of adequate stiffness, particularly lateral stiffness, is a major consideration in design of tall structures. The structure should be designed to limit lateral deflection to prevent II order P-delta effect due to gravity loading. With respect to serviceability criteria, deflection must be kept sufficiently low to allow proper functioning of non-structural components and avoid distress in structure. The structure must be sufficiently stiff to prevent dynamic motion becoming large enough to cause discomfort.

2.4 Human Comfort

It is very important to consider the human comfort criteria. A person living in the top storey should not feel uneasy due to drift. The building will sway to some extent when the building experiences horizontal forces. This is to ensure that the building is elastic. The design should make sure that the sway is limited to ensure that the people dwelling in top stories do not feel the discomfort both in terms of physical and psychological.

2.5 Fire

Fire protection is a very important factor with regards to a tall building. The approach to fire protection is quite different from that of a normal building. The distress due to fire can have cascading effect and can create heavy losses. The design of tall structure should involve factors which prevent fire. The design should also take care to ensure that in case of an accidental breakout of fire, the damage to the structure and the occupants is kept to a minimum.

2.6 Drift

Drift is defined as the ratio of maximum deflection at top of the building to the total height. The drift index can be used to estimate the lateral stiffness of the building. The general formula for control of drift is H/350 where H is height of the building. This value becomes very important in case there are two adjacent buildings close to each other. The gap between the two buildings should be greater than the value so that the force on one building does not affect the other in case of horizontal loading.

2.7 Foundation

Foundation of tall buildings should not only withstand the enormous gravity load coming from the building but all horizontal forces due to natural effects. Generally the foundation adopted for tall structures is either raft or pile or a combination of both depending the forces coming on foundation.

3. LOADS ON TALL STRUCTURES

Loads acting on tall structures are either directly by forces of nature or by artificial means created by humans. The two basic sources of loads are geographical and human-made. Geophysical and human-made forces are often mutually depended.

The geographical force, being the result of continuous change in nature, can be further divided as gravitational, meteorological and seismological forces. As a result of gravity the weight of the building itself produces forces on the structure called dead load, and this load remains constant throughout the lifespan of the structure. Meteorological loads vary with time and place and appear in form of wind, temperature, humidity, rain, snow, and ice. Seismological forces result from the erratic motion of the ground (i.e. earthquake).

The movements of occupants and equipment may cause the human-made sources of loading. They may also be the variation of shocks generated by cars, elevators, machines and so on or the result of blast and impact. Furthermore, forces may be locked into structures during manufacturing and construction processes.

4. COMPONENTS OF TALL STRUCTURES

The various components of a tall structure are Structural systems; Foundations; Energy Dissipation system; Cladding include materials used with respect to architectural and structural consideration; Mechanical and service systems such as heating, ventilation, air-conditioning, plumbing, and fire-protection; Electrical systems such as lighting, power distribution, grounding, lightening protection, emergency power, fire alarm system, and Vertical and horizontal transportation which include escalators and elevators for moving people and material.

4.1 Structural System

A structural system is an arrangement of structural components and a subsystem to properly transmit the vertical, horizontal and other forces acting on a building. Due to relative significance of structural costs for tall building construction, an improperly planned and designed system will substantially increase the total cost of construction. Furthermore, the system may not serve its purpose safely and properly.

A structural system is expected to:

- Carry dynamic and static load.
- Carry horizontal loads due to wind and earthquake effects.
- Resist stress caused by temperature and shrinkage effect.
- Resist external and internal blast and impact load.
- Resist and help damp vibration and fatigue effects.

4.1.1 Types of Structural Systems

4.1.1.1 Framed Structures

Framed structure is a type of structure having a combination of columns, beams and slabs. This is the most basic form of structural system for a tall structure. The various types of framed structures are:

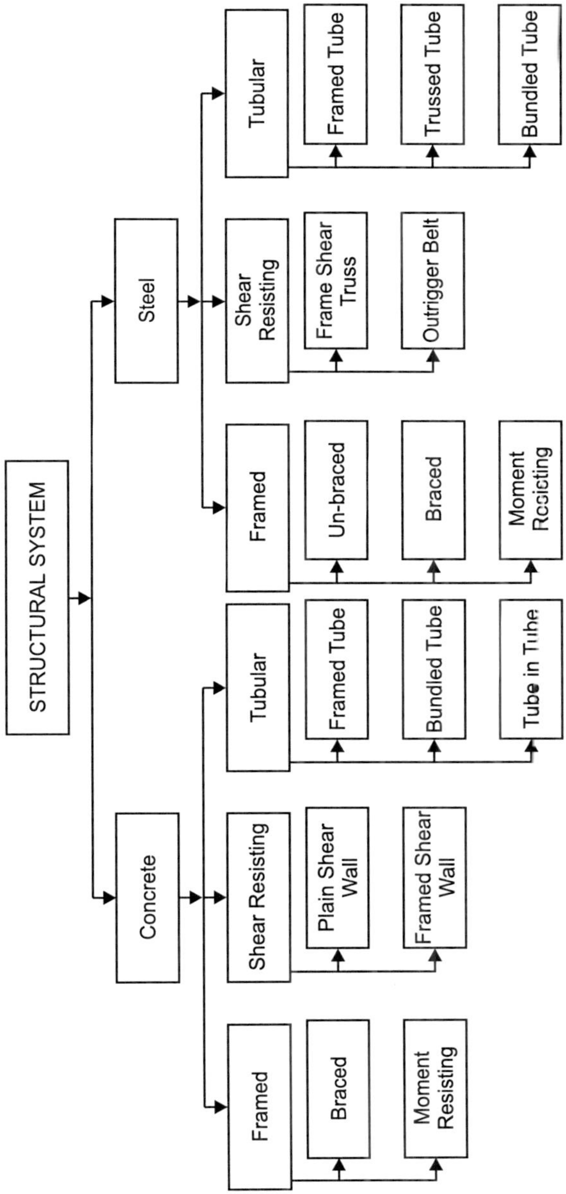

Fig. 9.1: Figure showing various structural systems

(a) **Un-braced Frames:** Un-braced frame is a type of structural frame which is designed to withstand the loads coming on the structure without any additional support members in the transverse direction.

(b) **Braced Frames:** Braced frames are like cantilevered vertical trusses with the fixity being taken at the foundation. They resist lateral loads primarily through the axial stiffness of the frame members. Axial shortening and the elongation of the column members under lateral loading accounts for 80 to 90% of the overall system deformation for slender truss system. The system is very effective in multi storeyed buildings of low to mid height range due to the high ratio of stiffness to material quantity.

(c) **Rigid Frames:** A rigid frame is a structural system which his like a skeleton constructed with straight or curved members interconnected by rigid connections. The structural system is termed as rigid frames because the joints in the frames are rigidly connected. The rigid connections resist movements induced at the joints of members.

(d) **Moment Resisting Frames:** They consist of linear, horizontal members (beams) in plane with and connected with vertical columns.

4.1.1.2 Shear Walled/Shear Resisting Structures

Frame action obtained by interaction of slabs, beams and columns are not sufficient to give required lateral stiffness for the buildings taller than 25 stories. Hence to give required stiffness, the buildings are provided with shear walls to function as main lateral load resisting element. Thus shear walls are vertical stiffening elements designed to resist lateral loads. They are usually continuous down to the base to which they are rigidly attached to form vertical cantilevers. These walls have very high inplane stiffness and strength, which makes them suited for buildings up to 50 stories. Shear walls may be a part of a service core or a stairwell or they may serve as partition between accommodations. The various types of shear resisting structures are:

(a) **Plain Shear Walled Structure:** Plain shear walled structures are the type of structure which consists only of shear walls.

These walls can be planar, open or closed or combination of these. There is no limitation to geometric configuration of shear wall system. Open systems are made up of single linear systems or a combination of such elements that do not completely enclose a geometric shape. Closed systems, in contrast, enclose a geometric shape and space. The common forms are squares, triangular, rectangular, and circular cores. The shear wall systems, whether inside or outside the building, may be arranged symmetrically or unsymmetrically. The shape and location of shear walls have significant effect on the structural behaviour under lateral loads, for example, a core eccentrically located with respect to the building shape has to carry torsion along with bending and direct shear.

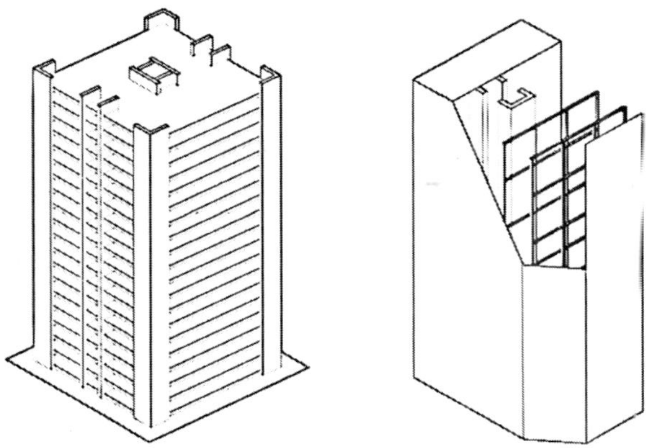

Fig. 9.2(a): Plain shear wall structure

Fig. 9.2(b): Framed shear wall structure

Fig. 9.2: Figure showing shear resisting system

(b) **Outrigger and Belt Truss System:** In belt truss system, also known as the core-outrigger system, the axial stiffness of the perimeter columns helps in increasing the resistance to overturning moments. This structural form consists of a central core, comprising either braced frames or shear walls, with horizontal cantilever—outrigger trusses or girders connected to the core of the outer columns. The core may be centrally located with outriggers extending on both sides or it may be located on one side of the building with outriggers extending to one side of the building.

(c) **Frame Shear Truss:** This type of structural system is used for structures made of steel. This is a combination of steel rigid frames and steel shear truss. The rigid frames take care of vertical loading and shear truss takes care of the horizontal loading.

4.1.1.3 Tubular Structures

The tube concept is efficient for tall slender buildings. In this system, the perimeter of the buildings consists of closely spaced columns connected by a relatively deep spandrel. The resulting system works as a giant vertical cantilever and is very efficient because of the large separation between the windward and leeward columns. The tube concept in itself does not guarantee that the system satisfies stiffness and vibration limitations. The chord drift caused by the axial displacement of the columns and the web drift brought about by the shear and bending deformations of the spandrels and columns may vary considerably depending upon the geometric and elastic property of the tube.

(a) **Framed Tube:** This is the simplest form of tube structure. In this system, the perimeter of the buildings consists of closely spaced columns connected by a relatively deep spandrel. The resulting system works as a giant vertical cantilever and is very efficient because of the large separation between the windward and leeward columns. The tube concept in itself

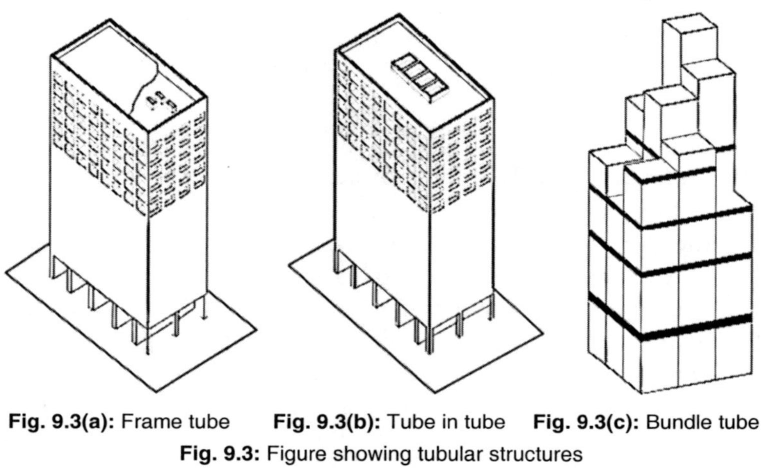

Fig. 9.3(a): Frame tube **Fig. 9.3(b):** Tube in tube **Fig. 9.3(c):** Bundle tube

Fig. 9.3: Figure showing tubular structures

does not guarantee that the system satisfies stiffness and vibration limitations. The chord drift caused by the axial displacement of the columns and the web drift brought about by the shear and bending deformations of the spandrels and columns may vary considerably depending upon the geometric and elastic property of the tube.

(b) **Exterior Diagonal Tubes:** Fazlur Khan proposed the idea of exterior diagonal tubular systems as early as 1972. He visualized the concrete version of diagonal truss tube system consisting of exterior columns spaced at about 10 ft center with blocked out windows at each floor to create a diagonal pattern on the facade. The diagonals could then be designed to carry the shear forces, thus eliminating bending in the tube columns and girders.

(c) **Tube in Tube:** This type of structural system is also known as hull and core system. In this structural system both the core and the outside are tubular structure. Generally the internal core is tube with all the services and utilities like elevator shaft. The external core takes care of the gravity and lateral loads.

(d) **Bundled Tube:** The bundled tube structure comprises of a set of tubes that are interconnected with common interior panels to form a perforated multicell tube as seen in plan. In this system, the frames in the lateral load direction resist the shear, while the flange frames carry most of the overturning moments. Each tube can be curtailed at different heights without diminishing structural energy. The closed form of the modules gives high resistance to the torsional loads. The larger spacing of the columns and shallower spandrels allows provisions for larger window openings compared to single-tube structure. A distinct advantage of bundled tube is that the individual tubes can be assembled in any configuration and terminated at any level without loss of structural integrity. This feature enables creation of setbacks with a variety of shapes and sizes. The disadvantage, however, is that the floors are divided into tight cells by a series of columns that run across the building width. The structural principle behind the modular concept is that the interior rows of columns and spandrels act as internal webs of a huge tubular cantilever in resisting shear forces, thus minimizing the shear lag effects.

4.2 Foundation

The structural design of a skyscraper foundation is primarily determined by loads transmitted by its many floors to the ground on which the building stands. To keep its balance in windstorms and earthquakes, its foundation requires special considerations because the lateral loads transferred to the soil, are rather large. Where load bearing rock or stable soils such as compact glacial tills are encountered at reasonable depth, the foundation may be directly carried down to the load bearing strata. This is accomplished by utilizing deep basements, caissons, or piles to carry the column loads down through poor soils to compact materials. The primary objective of the foundation system is to provide reasonable flexibility and freedom in architectural layout. It should be able to accommodate large variations in column loading and spacing without adversely affecting the structural system due to differential settlement. In places where thick deposits of compressible soil like clay is encountered, then the loads must be controlled to keep settlements to acceptable limits. Raft foundation can be adopted for a building of height of 25 stories.

4.3 Energy Dissipation System

All the energy that is induced in the structure by wind, blast earthquake or machinery is finally dissipated by damping forces. The damping, limits and progressively reduces the deformation and accelerations of the induce vibrations. The various ways to achieve damping are as follows:

4.3.1 Natural Damping

Structural damping is caused by friction developed within the particles of building materials. The damping forces are proportional to strain and deformation. A large amount of energy is also dissipated even for a minute slippage at the connection of members. The best example of this is the relatively high damping of bolted connection compared to welded steel structures. The increased friction of the particles in cracked concrete is also well known phenomenon.

4.3.2 Plasticity of Structural Material

Due to plasticity of materials, especially steel, energy is absorbed inelastically. The energy absorption by ductility is not limited to steel alone, but also occurs in combined material like reinforced concrete. The confining of concrete by reinforcement increases its ductility and capacity to absorb energy. These facts have led to ensure ductile detailing of structures, mainly in reinforced concrete especially where at joint and connection point high moments are generated.

4.3.3 Highly Absorbent Structural Systems

The structural systems are designed in such a manner that the damping system is present in the structural system itself. The structure in itself absorbs the energy without any external elements. The various ways in which this is achieved are:

(a) **Redundant Systems:** The capacity of the total structural system to absorb energy is largely influenced by its redundancy. This means that many redundant members participate in resisting loads. The overstressed portions of a structure can, after deformation, be relieved by neighboring members.

(b) **Combined Systems:** In the case of the use of combined structural system, for instance frames and shear walls, it is important for the total energy absorption that each system comes into action before any other experiences cause failure.

(c) **Bridged Systems:** In this system the two adjacent structures are connected by a bridge. Increased energy dissipation can be achieved by bridging structural systems. The increase in internal energy reduces deformation and acceleration precipitated by horizontal forces. This system changes the free standing cantilever of a tall building to a guided cantilever and the coupling of building restricts, orients and filters vibrations. Special devices provided in their linkage to dissipate energy can increase the effectiveness of damping.

4.3.4 Artificially Increasing Damping

The damping can be artificially increased by various means:

(a) **Provision of Viscous Damping by Coating, Joints, or Bearings:** Artificial viscous damping can be effective in reducing the amplitude of vibrations in buildings and building elements, bridges, machines and plates. This is achieved by absorbing energy in viscous coatings, joints, or bearing pads.

(b) **Use of Dynamic Vibration Absorbers:** Damped and undamped vibration absorbers are used to a large extent in any kind of machinery. However, the use of a large viscously damped absorber has been, until now, considered for only a few tall buildings.

(c) **Use of Coulomb Friction Joints:** The natural coulomb friction between steel plates or between steel and asbestos braking materials can be effectively used in friction joints. Depending on the movements and the face pressure, large amount of energy can be dissipated to form heat to damp the structure and reduce the deformation and acceleration induced by wind or earthquake. The joints can be provided in wind trussing of tall buildings, in cable connections of guyed towers, or externally attached to existing buildings.

4.3.5 Advanced Foundation Design

The use of special provisions in foundations such as spring supports, shock absorbers, and vibration dampers is well known in the design of machine foundations. Similar concept is applied in the design of building foundations, and some attempts have been made to reduce the energy input caused by earthquake in the structure.

The shock absorbing soft story concept has been advanced by Caspe. The isolation device prevents the foundation from transferring the full effects of the ground vibration in the substructure. The foundation, however, is able to resist all probable wind loading.

4.3.6 Aerodynamics Provisions

Though not frequently considered, it is a common knowledge that energy induced in the structure can be dissipated by aerodynamic damping. Certain aerodynamic shapes essentially reduce the energy induced in the buildings by wind.

4.4 Cladding

Cladding system is analogous to clothing a human body. It is an expensive part of a building taking up anything between 15 to 20 per cent of the structural cost. In tall buildings it is even more important because of the need for structural integrity and its resistance to various natural forces like wind, rain, sunlight, temperature and atmospheric pressure. The load coming on cladding is transferred to the structural system and hence it is necessary to design the cladding system in sync with the structural system and building use. The various types of cladding used are:

(a) **Conventional Building Material:** Earlier buildings used the conventional building materials like stone, bricks or blocks as a cladding material. These materials were easily available and cost effective. However as the height of the buildings started increasing, using these materials became expensive. These materials are heavy and due to this it increased the gravity load to be resisted by the structural system.

(b) **Toughened Glass or Curtain Wall Glazing:** Multi-storied buildings in early 20th century were constructed using steel. To reduce the gravity load glass came to be used as a popular material in facades. In the earlier days, the glass panels were fixed to the steel mullions. During the 1970s, the widespread use of aluminium extrusions as mullions made it easier to mount the glass and carry out the sealing of joints. The advantage of aluminium is, it is light weight and it can be moulded into complex shapes.

(c) **Composite or Sandwich Panels:** To further reduce the expenses on cladding, sandwich panels were introduced. A sandwich panel is a structure made of three layers: low density core inserted in between two relatively thin skin layers. This sandwich setup allows the achievement of excellent mechanical performance at minimal weight. The very high rigidity of a sandwich panel is achieved thanks to interaction of its components under flexural load applied to the panel: core takes the shear loads and creates a distance between the skins which take the in-plane stresses, one skin in tension, the other in compression. The most common sandwich panel is ACP (Aluminium Composite Panel). ACP is a type of panel which consists of two thin aluminium sheets bonded to a non-aluminium core. Aluminium sheets can be coated with polyvinyle fluoride or polyester paint. Apart from its structural advantages, easy maintenance has made it the most popular cladding material.

4.5 Mechanical Systems

The design and provision of mechanical systems are to be taken care of while designing tall buildings. Generally all the mechanical systems are provided in one floor and in a conventional building which is not too high, they are provided either in the basement or on the roof or both. In skyscrapers a mechanical floor is provided to cater to the needs of all services. Equipments like pumps, plants for HVAC etc. are positioned in this floor. This floor is accessed only by service elevators and due consideration is given for the loads during the structural design. On a lighter note, generally the 13th floor is used as service floor due to superstition.

4.5.1 Plumbing

Supply of water and provision of drainage is one of the most difficult aspects in a tall building. Due to its height and the volume of water required, many problems relating to pressure and delivery are encountered. Supply of water at adequate pressure to each occupant is critical. The two common methods in use are the gravity system and direct pumping system.

In gravity system there is a storage tank at the roof and the water is supplied to lower floors through gravity systems. The pressure required to ensure that the water reaches the overhead tank will be high. For a 40-storied building the pressure required to pump the water to the tope would be close to 25-bar. Providing mechanical system to pump at such a pressure and manufacturing pipes to withstand so much of pressure would be impractical. To control the pressure in pipes, pumps are provided at intermediate floors. These pumps maintain a constant pressure.

In direct pumping system, the water is pumped directly from ground upwards. Like gravity system, the pressure is maintained provided at intermediate floors. However in this case the intermediate pumps perform the role of increasing the pressure.

Waste disposal system in early days consisted merely of collecting waste water and sewage and dumping it to ground or a river. However the volume of waste generated and the world becoming more environmental conscious have made it necessary to look for systems to effectively deal with the waste. Recycling of water is also an important factor which is taken into account and provisions are made to recycle waste water.

One of the major issues faced in plumbing of drainage is the air pockets which will reduce the effective diameter of the pipes and create blockades. It is necessary to create ventilation systems so that this does not affect the flow of waste.

4.5.2 Heating Ventilating Air Conditioning (HVAC)

As the building goes higher, both the temperature and pressure reduce. This can provide discomfort to the occupants. HVAC systems are designed to ensure that the pressure and temperature inside the builds are within bearable limits of human comfort. Tall buildings also use great deal of energy and the space available to accommodate the mechanical systems is limited. The HVAC systems to be used should have the following criteria:

(a) **Custom Comfort:** Each living space needs customized control for individual comfort. The system should be capable of varying the amount of refrigerant being piped to individual air handlers. This customization also helps with the issues of varying demand between upper and lower floors.

(b) **Energy Efficiency:** The technology used should minimize energy consumption. The system should be designed to reuse heat given off in the condensing process to provide heat in other areas of the space, so it can be used for supplemental heating.

(c) **Quiet Operation:** The operation should not give out much noise and make it uncomfortable for occupants.

(d) **Less Space Requirement:** The system air handlers used should also translate to less indoor space required to house the units.

(e) **More Dependable:** The system used should undergo less wear and tear, which results in fewer breakdowns.

4.5.3 Fire Safety

Fire safety is one of the major factors, which has to be taken care of in tall buildings. The fire damage can be enormous if it is not prevented. Even when the fire breaks out sufficient provisions have to be made so that the fire is extinguished in time and damage is controlled. Measures should be taken so that adequate safety measures are provided. Following are the measures to be taken for fire safety and fire management:

- Isolated fire exits should be provided for faster and safer exits.
- Every floor should have a fire extinguisher and firefighting equipment which can be automatic as well as manual.
- A permanent dead storage of water for firefighting purpose alone should be provided. The dead storage of water at the top should be a minimum of 10% of the live storage at the bottom. Pipes should be provided throughout the height of the building and pressure head maintained at every floor.
- Fire fighting exercises should be carried out at regular intervals so that people don't panic when the fire breaks out.
- Means must be included in the building to assist in early detection of a fire to minimize its impact on both the life and property.
- The design of the building should consider compartmentalization to limit the spread of fire and early extinguishment. It should also include proper ventilation system for dispersion of smoke.
- A communication system is necessary not only to allow transmitting proper instructions to building occupants in case of a fire emergency, but also allow those people responsible for proper fire response to communicate with one another. This would involve not only internal two-way communication, but also quick notification to fire department.

- The elevator system should be designed in such a way that it can work in isolation during fire fighting. The means for this independent control should be included in the overall system of fire response.
- Management must include not only fire resistant values of walls, floors, shafts, and structure, but also of the fuel loading of the contents of the building (both furnishing and equipment) to minimize combustibility and potential smoke generation.

4.6 Electrical System

An electrical system is a network of electrical component deployed to generate, supply and transfer electrical power. The determination of electrical load for a tall building is one of the first tasks to be performed by the design electrical engineer. In addition to normal lighting and utility loads, this estimate should include capacity for building's mechanical systems, vertical transportation equipment, and any other special systems that are planned which require electrical energy, such as computers and their associated air conditioning systems and electrical heat. The load figure should also include spare capacity to satisfy the growth that might take place during the life of the building. Adequate safety systems should be provided in case of accidents due to electrical outages.

4.6.1 Power Generation

Once the electrical load has been estimated, for both initial and ultimate requirements, all the available service voltages offered by the various sources of electricity to the building have to be found out and arrived at. Generally the supply of power is taken from the main grid of the city. However, the power requirements generally tend to be very large for a tall building and the systems are created to generate power from renewable sources, for example, Petronas tower uses both wind as well as solar energy to generate power. There should be adequate provision for backup power generation.

4.6.2 Power Distribution

Once the electrical load has been estimated, for both initial and ultimate requirements, all the available service voltages offered by the various sources of electricity to the building have to be found out and arrived at.

The main service entrance for high voltages must be compatible with the utility company's distribution equipment. It is essential that all the main service-entrance and the associated distribution equipment for these higher voltages be properly coordinated with the utility and selected to achieve the utmost in safety and reliability. It should also permit future simplified maintenance, testing, and replacement.

The low voltage service equipment should be disconnectable from the power supply, either manually or automatically, without danger to the operating personnel. When automatic current breakers are provided as a service disconnect device, current limiting fuses properly coordinated with breaker tripping characteristics should be used in combination.

4.6.3 Lightning Protection

If a tall building constructed of structural steel is to have lightening protection system, common practice is to use the steel columns as the down conductors for the roof ground loop and lightening arrestors. All such installation should have these columns grounded at the column base by means of a ground electrode and have all electrode interconnected by means of a buried counterpoise cable to minimize the electrical impedence in the metallic conducting path of such grounded equipment. Consideration should be given to the installation of the ground loop conductor interconnecting the down conductor columns at selected floors throughout the height of the building.

A lightning protection system applied to a roof of the building of a RCC structure generally will require down conductors from the roof arrestors and roof ground loop to grounding electrodes connected to a counterpoise ground loop at the column footings. The reinforcing bars may be used as down conductors only if the bars are welded together and are of adequate size to achieve low impedance path. In a reinforced concrete building it is very important to ground all metallic equipment and piping to the ground loops or down conductors in order to provide a neutralizing path to the earth.

4.7 Horizontal and Vertical Transportation

Movement of traffic within a tall building is another important factor in the design of tall building. Transporting both human traffic and goods over a considerable height and distance require both speed and comfort.

The conventional elevators which move very slowly will not be practical in skyscrapers. Traffic needs to be transported within a reasonable time and hence you need to increase the speed. Due to the phenomenon of pressure drop as we go higher, there can be human discomfort while travelling within the elevator. The acceleration and deceleration of the elevators have to be gentle.

Escalators are provided between floors for movement of people. Escalators can be used for a limited number of stories as the travelling speed of escalator is slow.

Due to the height of the building, the skyscrapers have a large base. Travelling within the floor on foot can be tiring. For this purpose travelators are provided for horizontal transport. Travelators are similar to escalators but move horizontally.

Reference and Further Reads

Ross, Donald E., "HVAC Design guide for tall commercial buildings", 2004, American Society of Heating Refrigerating and Air-Conditioning Engineers.

Smith, Stafford and Alex Coull, "Tall Building Structures: Analysis and Design", 2011, Wiley India Pvt. Ltd.

Taranath, B.S., *Steel, concrete and composite design of tall building*, 2nd edition, 1997, McGraw Hill.

10 URBAN PLANNING

Urban planning is a process which involves both technical as well as political process concerned with the development and use of land. It involves preparation and passing of laws for permits for development, protection and use of the environment, public welfare and the urban environment including air, water, and the infrastructure passing into and out of urban areas, such as transportation, communication and distribution network. Urban planning is also referred to as urban and regional planning, town planning, city planning, rural planning or some combination in various areas worldwide. It takes many forms and it can share perspectives and practices with urban design.

Urban planning guides the development of an area or town or a city. Though the term urban planning looks more of a term which deals with city or town, it also takes into account the development of rural areas. Although predominantly concerned with the planning of settlements and communities, urban planning is also responsible for the planning and development of water use and resources, rural and agricultural land, parks and conserving areas of natural environmental significance.

Early urban planners were often group of members who were landscape architects, civil engineers and public administrators. Today urban planning is a separate, independent professional discipline.

The concept of urban planning is found in ancient civilizations. Archeologists have found evidences of planned cities in the ruins of early civilizations. Archeological excavations and discovery of cities like Mohenjo-Daro and Harappa give enough evidences of the presence of urban planning and design in ancient civilizations. Hippodamus, called by Aristotle as the father of city planning, is considered the inventor of orthogonal urban layout.

1. URBAN DESIGN

The difference between urban planners and urban designers is very distinct. Urban planners work on policies that shape the urban development and urban designers give a physical form to the development taking into account the parameters of the policies laid down in planning.

Though the term urban design is relatively new, the concept of urban design is as old as civilization. Ancient and medieval cities were meticulously planned. The design of cities and towns reflected the mindset and living conditions of the prevailing society. Twentieth century industrial revolution brought about a major shift in the field of planning and architecture that set the foundation for modern urban design. Sir Ebenezer Howard's idea of garden city looked at creating a balance between ecology and development. The increase of automobiles over the years had a major impact on the field of urban design. Design of spaces had to revolve around the roadways and traffic patterns.

The current trend of urban design looks to improve on Sir Ebenzer's idea by creating sustainable environment and over all livability. It aims to create smart cities which reduce the environmental impact, support sustainable transport and create ecological balance.

Urban design looks into the following aspects to name a few:

- Urban structure
- Urban typology, density and sustainability
- Ecological and cultural balance
- Transportation including accessibility and pedestrian movements
- Aesthetics
- Interaction between utilities.

2. SPACE PLANNING OF BUILDINGS

Though the subject of space planning is more of an architectural subject, it is important for engineers to have the knowledge of planning rules and bylaws. Civil engineer does play an important role in planning as the role of licensed surveyor is extended to civil engineers too. The spatial planning should not just be for aesthetics but also take into consideration the practical aspects of living. Space planning of building involves 2 parts, namely External Planning and Internal Planning.

2.1 External Planning

External planning is related to the design of spaces outside the building. The national building code specifies the criteria for the spaces to be left around the building not only from the consideration of natural lighting, ventilation etc. but also from the consideration of structural stability. The various considerations in external planning are as below:

2.1.1 Means of Access

Every property needs to be provided with proper means of access. This is generally called the road or a passage as the case may be. The roads can be public roads or private passages. The roads should be such that they ensure the flow of traffic is smooth and seamless. They should not have any blockages and should be straight. The width of means of access is dependent on the use of the property and building and the distance from the local road. The minimum width required is 6 m. Number of dwellings and the height of the building are also dependent on the road width.

2.1.2 Minimum Land Area and Size

The development of each plot for various uses has its restrictions based on the requirements and occupancy of the development. For each type of building there is a minimum land area specified. This enables the optimum design and utility for the desired function the property is developed, for example, for a residential plot in Chennai, Tamil Nadu the minimum land area required is 90 m^2 and the minimum frontage required is 6 m.

The extent of plot on which a building can be built is called the plot coverage. The restriction on plot coverage also ensures the spaces around the building. This too varies depending on the type of building. For a residential building in Chennai, the plot coverage is 65%.

The restriction on the height of building is kept taking into consideration the surroundings and the serviceability. The Floor area ratio (FAR) or Floor space index (Index) also decides the height of the building. FAR is the total floor area with respect to the land area. Another criterion which determines the height of the building is the provision of fire fighting capabilities.

2.1.3 Set Back Spaces

The spaces around the buildings should be such that the sunlight and air circulation is not affected and the building receives enough natural light and air. The distances between the buildings should ensure that the forces from one building do not affect the stability of the adjacent buildings.

The space around the building is termed as setback space. The setback spaces vary depending on the size, height and the utility of the building. Normally the setback space for a two-storied residential building is 1.5 mts on either side. This is specified to ensure that there is enough lighting and ventilation for the building.

From the structural point of view, generally the size of foundation for such a structure is around 1.5 m × 1.5 m. The effect of the load distribution on the soil from the foundation is taken at an angle of 45 degrees from the edge of footing. It is to be ensured that the angles don't intersect and there are no stresses induced to the soil below the foundation due to the adjacent foundation.

The access for providing fire fighting also determines the setback spaces. For multi-storied and public building the setback spaces are more to provide access for fire engines around the building. A fire engine should be able to go around the building.

2.1.4 Landscaping and Water Conservation

Landscaping has an important role to play in the plot or land development. It is important to ensure that any development is in synergy with nature and enhances environment quality. Greenery should be provided to ensure the stability of atmosphere. However care should be taken to ensure that the landscaping does not affect the stability of the structure and also provides unrestricted natural light. Trees that have horizontally expanding roots should be avoided closer to the structure.

With potable water becoming a critical issue in modern day scenario, it is important to conserve water. Natural water bodies should be protected by ensuring that no structure permanent or otherwise is developed within a certain distance. The recharge of wells and aquifers should be done by carrying out rainwater harvesting. Recycling of water should be done wherever possible.

2.2 Internal Planning

Internal planning relates to the design of spaces within a building. The internal planning should consider the comfortable living and movements of the occupants. The space planning should also ensure that the air circulation is optimum and rooms don't become claustrophobic. The design of the openings of building should also be such that enough natural light and ventilation is received for the dwelling space.

In the olden days, the internal spaces and the sizes of the room was based on the limbs of the human occupants. There were no standard units of measurements. The design of spaces was such that the occupants could move within the spaces with minimum effort. Researches indicate that the perception of dimensions is better if it is related to the sizes of human form. Leonardo da Vinci's depiction of vitruvian man demonstrates the basis of co-relation between human limbs and dimensions.

The attempt to standardize the measurements and the introduction of metric system put an end to the human perception of spaces. Though it made the work of the designer and the executers easier, it poses a problem in the human imagination of spaces. Even though designers try to create an accurate mental picture, more often than not, it gives a false idea of the rooms and the surprises the occupants once the work is actually implemented.

Design of internal spaces is not just a physical requirement of human but also an emotional need of human mind. Every individual feels differently and has unique requirement of spaces. Some may prefer larger rooms whereas others may prefer a more optimally sized room. The design of spaces with respect to division, utility, décor, lighting etc. is more of an individual choices rather than a generalized one.

Vaasthu Shastra, the ancient concept of space planning, very specifically talks about the holistic design of space with respect to its occupant. The system of measurement described in vaasthu is called tala system and does not depend on the absolute dimensions. It defines all dimensions as a relationship of proportions of component parts with respect to the whole.

The concept of apartments has given space planning in a totally new dimension. Customization for large numbers became an impossible task and designers looked at standardization of the sizes of the dwellings. The optimization of sizes were arrived at considering the various aspects, such as requirements for moving and furnishing, air circulation, natural lighting etc.

In India, the minimum room sizes and ceiling height has been specified by the NBC (National Building Code). Some of the points to be followed in the design of the rooms are:

- The height of habitable room in a residential should be a minimum of 2.75 m and in a public building should be 3.6 m.
- The size of a habitable room should not be less than 9.5 m^2 with a minimum width of 2.4 m. This means that the minimum size of the room should be 2.4 m × 3.96 m.
- The area of kitchen should not be less than 5 m^2.
- The area of the bathroom should not me less than 1.8 m^2 with a minimum width of 1.2 m.
- The minimum clear width of the staircase for a residential building is 1.0 m. The minimum width of tread should be 250 mm.

- The openings for windows and ventilators are dependent on the climatic condition. For a dry hot climate, the area of openings should not be less than 10% of the floor area of the room.

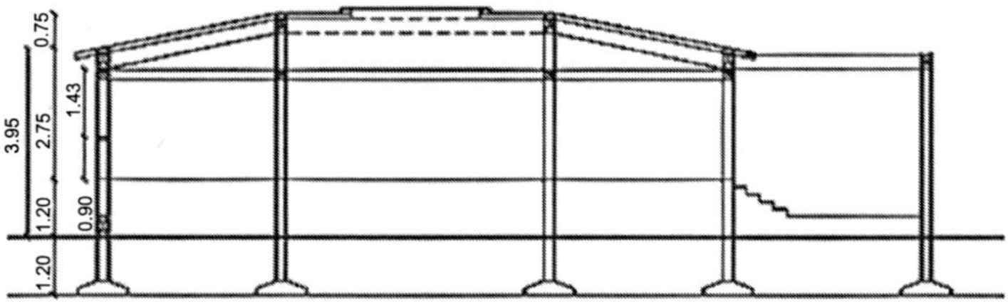

Fig. 10.1(a): Figure showing frame structure in section

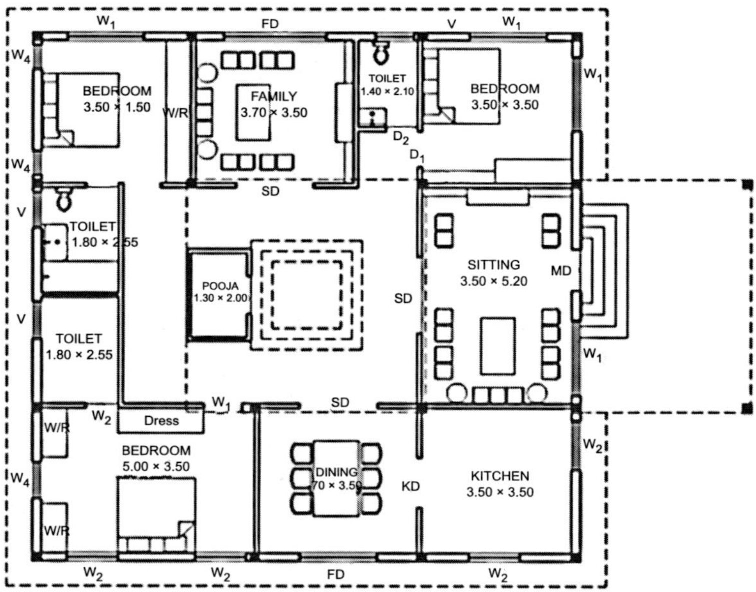

Fig. 10.1(b): Figure showing plan

Fig. 10.1: Figure showing plan and section of a house

3. GREEN BUILDING

Green building (also known as green construction or sustainable building) refers to both a structure and the using of processes that are environmentally responsible and resource efficient throughout a building's life-cycle. The concept of green construction starts from design conceptualization, to construction, to operation, maintenance and renovation. Even in demolition the process needs to look into the effect on environment it can have. In other words, green building design involves finding the balance between human made structures and the sustainable environment. This requires close cooperation of the design team, the architects, the engineers, and the client at all project stages. The Green Building practice expands and complements the classical building design concerns of economy, utility, durability, and comfort.

US green building council developed a set of rules and rating system to define and categorize green buildings. The certification is called LEED (Leadership in Energy and Environmental Design). The common objective of green buildings is to reduce the overall impact of the built environment on human health and the natural environment by:

- Efficiently using energy, water, and other resources.
- Protecting occupant health and improving employee productivity.
- Reducing waste, pollution and environmental degradation.

Although new technologies are constantly being developed to complement current practices in creating greener structures, the concept of green building is not new. The design and construction of buildings in earlier days focused on using natural locally available materials and sustainable design to ensure that the buildings were in synergy with nature. The construction of structures, especially houses were done using locally available materials. The construction materials were made from raw materials available within the periphery of the land, where the building was being constructed.

4. TEMPLE ENGINEERING

Contrary to the popular belief that the temples are just a place of worship, in South India especially in Tamil Nadu, the temples were the centre of town planning. The temples had far greater purpose and utility than being just a building for religious gathering. There was a scientific reasoning for all the activities and they had a practical application and benefit to the society at large. Everything from choosing the material for construction to the activities followed after the construction have been carefully thought of and followed in those days.

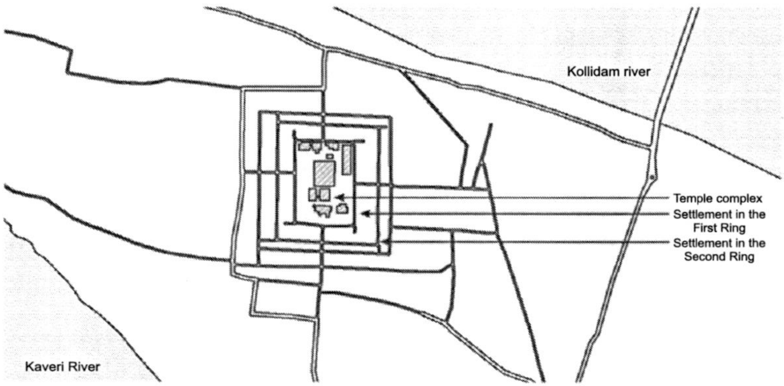

Fig. 10.2: Figure showing schematic layout of temple settlement (Srirangam, Tamil Nadu-India)

People generally look at temple from the architectural point of view. It is common to admire the aesthetics and beauty of the form of the temple and the sculptures carved in them. Very rarely people study the engineering aspects of the temple. It will be of great benefit to the civil engineers to study the engineering technology in the design and construction of a temple. From the design to the method of construction, it is quite evident that the people in those days had advanced knowledge of architecture, astronomy, structural engineering, construction technology, material sciences.

4.1 Utility of Temple

As mentioned above, the temple had far greater utility than just a place of worship. It served many purposes with respect to culture, history, urban planning and design and activities of the prevalent society. Following were the utilities of the temple:

4.1.1 Center of Town Planning

The temples in earlier days were the centre of town planning. It is believed that the architect/town planner/engineer in the king's court when asked to reconnaissance the area for a development of new town would find the spot where he will feel the maximum positive energy. That point was marked as the point where the sanctum sanctorum would be placed namely the position of the main deity. From there he would walk to all four directions till such point where he would feel the positive energy. The place where the effect of the positive energy ended was taken as the edge of the town. All the activities of negative energy such as crematorium were carried outside the town. The settlements or dwellings would be positioned in such a way that the people of similar activities would be grouped together and ensured that their travel to their workplace was easy. For example, the people working in temple would live close to the temple. The traders would have their dwellings set up closer to the market place and so forth. This is a perfect example of urban planning and design.

4.1.2 Community Hall

The temples were also a form of community hall where various activities were carried out. The sizes of the temples were large and hence the temple could hold large gatherings. The examples of some activities were weddings and temple festivals. Temple was and still is a place for socializing. Temple festivals were good place for showcasing of artistic performances such as dance and music. It was also a form of meditation centre.

4.1.3 Treasury

Quite a few temples also acted as a treasury. There was a safe room built in the temple which would hold the state's treasures. The deities were adorned with precious jewels, which in a way became the state property and it was under the safe keep inside the temples. The safe room was built in such a way that most did not know that it was a safe. The walls around the temple and the security were so designed, that they were practically impregnable. The famous temple of Somnath, which was invaded by Mahmud of Ghazni several times and the Padmanabha temple in Kerala stand testimony to this.

4.1.4 Fortress Against Calamity

The temple was a fortress against calamity both natural and man-made. The outer walls of the temple were designed and built to withstand any form of natural force experienced in the locality. The materials used were such too, that they could withstand constant weathering. The temple walls were also, almost impregnable and hence during war, the women, old people and the children were housed in the temple campus for safety.

4.1.5 Lightning Conductor

It was a rule that no structure in a town or a city could be taller than the temple. The dome of the temple was the highest point in the town. The main reason for that was the dome was also the lighting conductor and an earthing system. This made the sanctum sanctorum the most energized point in the entire structure.

4.1.6 Motel/Inn for Travellers

In the earlier days there were categories of travelers, those who visited their relatives, traders and pilgrims. Majority of the travelers those days were the sages who would travel to visit various religious centers. There were no designated hotels or restaurants for travelers. Every house had a porch, where travelers could spend the night or rest for some time. The temples were predominantly served as restaurants for travelers.

4.1.7 Granary

Some important temples were used as granary too. The dome of the temple also had an important function with regards to granary. The seeds of grains were stored in the dome of the temple. This was done ensure that in case of natural calamity striking the town and destroying all the crops, there were seeds to replant the crops. The seeds were replenished every twelve years.

4.1.8 Water Harvesting and Preservation

Temple ponds are a great example of water preserving and rainwater harvesting systems. The systems were so designed that all the potable water was directed to be collected at the temple pond. Efforts were made to ensure that the temple pond remained filled all the time. There are some temples where the temple pond is connected to the river nearby. The temple tank in Naganathaswamy temple in Thirunageswaram is connected to Cauvery river in Kumbakkonam, which is 5 kms away, via an underground aqueduct.

4.1.9 Museum and Library

The temples as served as a museum and a library. They displayed the culture of the era in which the temple was built. The sculptures and idols showcased the talents of the artists. The scriptures were inscribed on the walls to preserve the language and literature. The scriptures described stories and events related to the place and time and moral lesson they are supposed to propagate.

4.2 Design and Construction of Temples

The design and construction of temple has both been a fascination and mystery to the present day designers. The archeologists and historians have been trying to find answers of how, why and what of the activities carried out during construction. They still wonder, with the kind of equipments they have in those days, how they could achieve such tremendous feats. One classic example is that of Braghadeeshwara temple in Tanjavur. The dome of the temple is a single piece of granite and people are still wondering how the stone was placed on top.

As mentioned in 2.2 of this chapter, the design of temple was based on vaasthu. Vaasthu Shastra texts are instruction manual for practicing architects. They describe in detail the construction of temple from the selection of site, measurements and proportions, choice of materials and structural process to the establishment and consecration of presiding deity. The scope of vaasthu does not limit itself to architecture but extends to sculptures and paintings.

(a) **Material Choice:** In the ancient times, temples were the only permanent structures built. They were built to last for centuries and hence the material used for construction was so chosen that they withstood the climatic changes over long period of time. The choice of material had to be such that it was hard, strong and can withstand variances in climate over long periods.

The primary material used for construction was stone or granite. Stone is a naturally available material which has been seasoned. Granite being a hard material has a good resistance to weathering. Lime mortar was used as an adhesive between joints of the blocks of stone.

The flooring inside the main structure of the temple was done using granite. Generally it was the same material used to construct the structure. The flooring around the temple, on the outside, was done using locally available material. In most cases it was left as natural ground. This allowed the absorption and dissipation of rain and heat. In some places clay bricks were used as pathway around the temple. This also acted as a plinth protection.

(b) **Design:** The design of the temple was inclusive of both aesthetics and stability. Every element carved out of natural materials showcased the artistic nature of the designers without compromising on the strength and stability. As mentioned earlier, the designers looked into all the aspects of the utility of the structure with respect to present and future use.

The designers had good knowledge of astronomy. Each temple had its own angle of orientation and not always true north or east. The angle of orientation depended on the altitude and the declination of the place. Importance was given to the angle and time of sunrise at various time of the year.

As mentioned earlier, the internal space planning was done based on the size of the occupant and the presiding deity is considered the main occupant of the temple. Consequently the dimensions of sanctum sanctorum as well as the temple were taken with relation to the size of the main idol.

The structural design of the temple was kept simple. Design ensured symmetry to take care of overall structural stability. The effects of horizontal forces were minimal compared the large gravity load imposed by the self weight of the structure. The foundation was generally open foundation like a strip footing which runs along the wall. The depth of the foundation considered all the relevant factors like bearing capacity and overturning due to wind load. Till 2 decades back, it was considered that there would be hardly any earthquakes.

Fig. 10.3(a): Figure showing structure of Temple

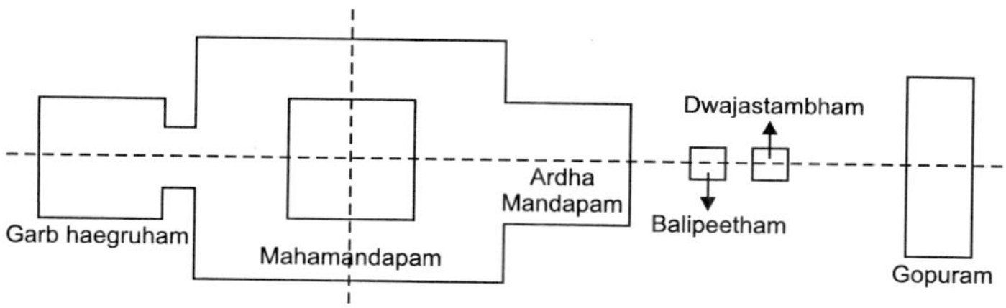

Fig. 10.3(b): Figure showing schematic plan of Temple

Fig. 10.3: Figure showing schematic plan and structure of Temple

The columns were designed as short columns. The columns were placed above the plinth level. The beams and slabs were designed as simply supported. The deflection criteria were kept in mind while designing the spacing of columns and the span of the horizontal members. The sizes of the members were determined to ensure that the structural members did not experience any tensile force.

(c) **Construction Techniques:** The mysteries of construction process adopted in those days are still being investigated especially for transport of materials such as large blocks of stone. Looking at the size of structures, it can be inferred that it was both time consuming and labour intensive. The builders of those days used to simple techniques of physics and mechanics. The knowledge of lever was very common and the materials were transported using a system of levers and pulleys.

The time was not a criterion in the completion of construction and the importance was given to the quality of construction and precision. Each element erected had a very specific purpose and it was important to be precise in the erection of the elements of the structure. It was important to maintain the accuracy of the dimensions and care was taken to ensure the same.

To conclude it would be appropriate to iterate the fact that, temples are also an engineering marvel and it included all the aspects of a development of a structure and a town. Be it urban planning and design or construction of sustainable structures, it would enormously benefit the present society to look into the engineering aspect of temples in great detail.

Reference and Further Reads

Ernst and Peter Neufert, "Architect's Data", 3rd Edition, 2002, Wiley-Blackwell.

Friedman, David, "Florentine New Towns: Urban Design in the Late Middle Ages", 1988, MIT Press.

National Building Code of India, 2005, 2nd revision, Bureau of Indian Standard.

Trivedi, Prof Kirti, "Hindu Temples—Models of fractal universe", 1993, Paper presented at the International seminar on Mayonic Science and Technology, Chennai.